Nuclear Power, Energy and the Environment

Books by the author:

The Optical Model of Elastic Scattering, 1963

Nuclear Reactions and Nuclear Structure, 1971

Nuclear Heavy-Ion Reactions, 1978

Growth Points in Nuclear Physics, (3 vols) 1980, 1981

Nucleon Momentum and Density Distributions in Nuclei, 1988
        (with A. N. Antonov and I. Zh. Petkov)

Pre-Equilibrium Nuclear Reactions, 1992
        (with E. Gadioli)

Nucleon Correlations in Nuclei, 1993
        (with A. N. Antonov and I. Zh. Petkov)

The Nuclear Optical Model, 1994

Introductory Nuclear Physics, 1997
        (with E. Gadioli and E. Gadioli Erba)

****

Spacetime and Electromagnetism, 1990
        (with J. R. Lucas)

****

Nuclear Physics in Peace and War, 1961

Our Nuclear Future? 1983

Energy and Environment, 1997

Nuclear Power, Energy and the Environment

Peter E. Hodgson

University of Oxford

Imperial College Press

ICP

Published by

Imperial College Press
203 Electrical Engineering Building
Imperial College
London SW7 2BT

Distributed by

World Scientific Publishing Co. Pte. Ltd.
P O Box 128, Farrer Road, Singapore 912805
USA office: Suite 1B, 1060 Main Street, River Edge, NJ 07661
UK office: 57 Shelton Street, Covent Garden, London WC2H 9HE

British Library Cataloguing-in-Publication Data
A catalogue record for this book is available from the British Library.

NUCLEAR POWER, ENERGY AND THE ENVIRONMENT

ISBN   1-86094-088-9
ISBN   1-86094-101-X (pbk)

This book is printed on acid-free paper.

Printed in Singapore by Uto-Print

PREFACE

This book is concerned with the problem of producing the energy that we all need to sustain our living standards without at the same time harming the environment that makes life possible and worth living. There is no shortage of energy. There is enough coal to last hundreds of years, and the sun pours onto the earth far more energy than we could ever need. The problem is to make it available in useable form without devastating the earth.

To tackle this problem we have to consider all possible ways of obtaining energy and then see how they affect the environment. We have to take into consideration their capacities, their costs, their reliabilities and their safeties, because these considerations determine whether they will be accepted or not.

Nuclear power is included in the title because it is the most promising and at the same time the most controversial of all the energy sources, and evokes strong opposition. Its association with nuclear weapons, the hazards of nuclear radiations, the problem of the disposal of radioactive waste, the threat of nuclear accidents and the possibility of diversion of plutonium to make weapons all combine to produce such revulsion that nuclear power is totally unacceptable to many people. It is nevertheless desirable to attempt an objective survey of the advantages and disadvantages of nuclear power compared with other energy sources. Other energy sources are also studied, but the main question is whether nuclear power should be phased out as soon as possible or whether it is set to be the energy source of the future.

Other energy sources such as wind and solar are also controversial, with their optimistic supporters. It is not at all easy to evaluate the conflicting claims made and to decide what is the best energy policy. As I am myself a nuclear physicist, I am readily suspected of being pro-nuclear, and indeed in a sense this is true in that I have a keen appreciation of its potentialities. This does not, however, prevent me from describing its disadvantages as well as I can. For example, I remain somewhat sceptical about the reality of global warming, although if it is established it provides a strong argument in favour of nuclear power. We are all inevitably influenced by our experiences, and mine include fifty years of research in theoretical and experimental nuclear physics, together with teaching physics in the universities of London, Reading and Oxford. For a long time I have been concerned about the effects of nuclear physics on our society, and was a member of the Atomic Scientists' Association for many years, serving on its Council from 1952–9 and editing its Journal from 1953–5. I have been a member of the Pugwash Conference on Science and World Affairs since its inception in 1955 and have written several books on the impact of nuclear physics on society including *Nuclear Physics in Peace and War* (Burns and Oates, 1961), *Our Nuclear Future?* (Christian Journals, 1983) and *Energy and Environment* (Bowerdean, 1997). I have drawn on my own experiences while participating in what is optimistically called the nuclear debate, particularly in the chapter on the political aspects of the energy crisis.

Like everyone else I obtain information from books and articles, but before accepting it as reliable I can ensure that the writers are well-qualified and are genuinely trying to establish the truth. I can also test what they write by the normal requirements of consistency, both internal and external, and also with the laws of physics. Being well aware that there are pressure groups that seek to influence public opinion by energetic propaganda campaigns makes it easier to assess their writings at their true worth.

One difficulty in attaining a balanced view of these complex problems is that most of the literature is strongly biassed in one way or another. It is extremely easy to make a plausible case for or against a particular energy source by stressing its advantages or its dangers and ignoring all contrary considerations. The literature published by the various energy associations tend to be over-optimistic about the advantages of their particular form of energy, blind to its disadvantages and overly critical of the rival sources. Most of the writers on energy are employed by industry or by trade or environmental associations, and this is often (usually quite unjustly) held to affect their impartiality. The only people who can speak freely without these constraints are independent writers and academics, and so we have a correspondingly great responsibility to develop and share our knowledge. This freedom from constraint is a great advantage, though inevitably it does not remove the handicap of lack of knowledge and experience that afflicts us all.

Quite apart from these difficulties, many of the situations discussed are changing more or less rapidly. The estimate of future population growth or of the reserves of various types of fossil fuel, to take just two examples, are all subject to inherent uncertainties. With the best will in the world, it is not possible to do more than to present what seems to be the best estimate, and to keep it under constant revision in the light of new developments. It is thus very easy for the discussion to become outdated, and so it is necessary to keep it constantly under review.

My aim is not to argue for a predetermined conclusion but to present the facts as objectively as I can and then leave the reader to draw his or her conclusions. I certainly have no easy solutions to the world's problems, and I doubt if there are any. But what is important is to base the energy debate on the acceptance of ascertained facts together with an estimate of the penumbra of uncertainty that inevitably attends any quantitative evaluation.

What is a cause of acute concern is that Governments are often forced by the pressure of (misguided) public opinion to take decisions that they know are not in the public interest. This is why it is vitally important that the debate on these problems should be based on accurate factual knowledge. Even with the best knowledge it is seldom clear what is the best course to take, but this is certainly better than taking decisions on the basis of ignorance and prejudice.

It is extremely important in all these discussions to express energy resources and needs, production capacities, costs and safety in a quantitative way, and to give some indication of the reliability of the numbers quoted. I have tried to do this as far as I can, but it is likely that some of those given are now superseded. To some extent this is inevitable in a changing world situation, and I would be grateful for corrections and improved data. Furthermore, it is not easy to avoid the uncertainties introduced by the advocates of some forms of energy who deflate their own costs and inflate those of rival sources.

The broad conclusion of the discussion of energy supplies is that providing the population growth is moderated we should have enough energy worldwide to provide our basic needs. A much more difficult question is whether we can do this without polluting our environment with serious detriment to our quality of life. This is the real question, and it should strongly influence our choice of energy sources.

I thank all who have, over the years, worked to improve the quality of the scientific information available to the public, in particular Professor Joseph Rotblat, who has been a continuing inspiration for many years, and indeed all my colleagues and friends in the Atomic Scientists' Association and the Pugwash Conference on Science and World Affairs. The publications of these bodies have been a most valuable source of information. Finally, I thank Mr. S. Sakurai of Kinseito Ltd., Publishers, and Mr. R. Dudley of Bowerdean Publishing Company Ltd. for kindly permitting me to use material from my booklet "Energy and Environment", and also those who have permitted the reproduction of figures.

P. E. Hodgson

CONTENTS

INTRODUCTION

All over the world people are realising more and more clearly that the increase of population and the growth of industries are seriously threatening the environment. The more people there are, the more energy will be needed, and this means more factories, more power stations, and therefore more pollution. We are gradually poisoning the earth, and if it dies, then we all die. We just cannot go on like this. If we are to survive, there are limits to the growth of population and living standards, and if these are forgotten our planet will be poisoned forever. In addition there is the possibility of world-wide climate change, with effects that are difficult to foresee.

This book is concerned with all these problems that are of vital importance for everyone. We are all dependent on energy to cook our food, to make our clothes, to light and heat our homes, to drive our transport and communications and to provide power for our factories. How best can we obtain the energy we need, with the least damage to the environment? There are some energy sources that have been used since time immemorial, while others have been developed only in this century. All consume the resources of the earth to a greater or lesser extent and all affect our surroundings, the air we breathe, the water we drink and the food we eat. Can we obtain all the energy we need without polluting the environment to an unacceptable extent? And if not, can we reduce our energy consumption sufficiently to avoid catastrophe?

An attempt to answer these and related questions raises many scientific, technological, sociological, political and moral questions to which there are no easy answers. For example, how much of the earth's resources should we use for ourselves, regardless of the possible needs of future generations? Should we go on producing energy in ways that pollute the atmosphere? Should industrial companies be allowed to devastate the landscape? These and many similar questions are inter-related in many ways that differ from one country

1

to another. Most of them raise moral questions that have been insufficiently studied.

Considered simply as an academic exercise, these are very difficult and complicated problems. Any attempt to solve them in an objective way is however made far more difficult by the reactions of those likely to be affected by any decision. Manufacturing companies naturally want to concentrate on making and marketing their product with maximum profitability, and until recent years have not been much concerned about the effects of their activities on the environment. If it is proposed to establish a repository for noxious waste, no one wants it to be sited near their home, whatever assurances they receive concerning its safety. Increasingly, all such activities are regulated by laws, and public enquiries are often held to try to find the best course of action. Inevitably these are confrontational and lead to fierce controversies, invective and propaganda. It is usually easy to make an apparently impressive case both for and against any proposed development. All this makes it more than usually difficult to reach an objective judgement.

In such circumstances it is more than ever necessary to make clear from the start the principles and methods that will be used to try to answer these questions. It is not of course possible to give complete answers: there are many areas of legitimate disagreement and other areas where the best course of action can only be decided when more facts are available. It will become clear, however, that there are other areas where enough is known to show that some proposed solutions are certainly impracticable. The acceptance of such areas of established knowledge is a precondition for responsible discussion in the areas of legitimate uncertainty. Unfortunately, however, the force of propaganda has largely prevented the attainment of consensus in the areas of established knowledge so that much contemporary discussion is worse than useless. This naturally has serious implications for the future of our society, as it makes it much more difficult for Governments to reach responsible decisions.

How can it be ensured that the discussion of these problems takes place realistically, respecting what is already known? We can begin by accepting, at least provisionally, what has been written by real experts on each subject, that is people who have the necessary qualifications and have devoted years of study to the question. Especially useful are the reports of national and international committees of experts. This is a good beginning but it is not enough. Such documents often deal with only one or a few aspects of a problem or are concerned with a limited geographical area. They inevitably become dated, and are not immune from political influence. We also have to be aware of the existence of pseudo-experts, that is people with sufficient scientific knowledge to give their writings a spurious authority, who pose as energy consultants and write very plausible but politically-biassed articles for pressure groups who use them to enhance the credibility of their propaganda.

To assess and relate these studies we can apply the criteria of internal consistency and conformity with known physical laws. This may indeed seem obvious but it is a common experience to find proposed solutions that conspicuously fail to satisfy this requirement. The importance of quantitative analyses cannot be sufficiently emphasised. Whenever any solution is proposed, it is essential to insist on numerical data, supported by statistical analysis. Without numbers, it is impossible to assess the importance of competing factors. Very often one finds general statements, based on a few events or qualitative data, making an apparently impressive case that collapses as soon as some numbers are introduced. The numerical data used may not be very accurate, but providing we have some measure of the likely uncertainty it is infinitely better than no data at all. Without numerical data, vital arguments are at the mercy of emotional propaganda. For example, we are exposed to many hazards in our everyday lives. Studies have shown, however, that people are often concerned about statistically negligible risks, while thoughtlessly exposing themselves to relatively serious risks.

Much of the discussion concerns what we should do now to bring about some desired result, or to avert some calamity, in the future. We are thus inevitably drawn into trying to predict what is likely to happen on the basis of our present knowledge, and such extrapolations are fraught with hazards. What will be the population of the world in fifty years' time? What will be the energy demand then? Will the future climate changes continue to follow the trends of the last few decades? Will some new energy source be discovered? No one knows the answers for certain, so we have to make the best estimates that we can. Many such estimates are quoted, and there are inevitably some inconsistencies among them, and these have been allowed to stand as a measure of the existing uncertainties.

A further difficulty is that in most cases the data required for a well-based decision do not exist, but if we wait until we have enough data it will be too late. Events will not wait for us; we have to decide now. In some cases it may already be too late. It may sound very reasonable and responsible to say that we must wait until we have enough evidence to be sure that we can make the correct decision, but in reality this is a totally unreasonable and irresponsible way to behave. This is not an unfamiliar situation: most of our lives we have to decide what to do now on the basis of incomplete knowledge, and then live with the consequences. The option of doing nothing is often the most dangerous of all.

It must also be realised that not all problems have unique solutions. Energy can be provided in a number of ways, each with several advantages and disadvantages that have to be balanced against each other. The analogy of balance, however, is inappropriate because they are usually incommensurable. How can one balance cost against safety, or reliability against effects on the environment? This is one of the areas of legitimate debate, and people with different priorities may come to different conclusions. In addition, the energy needs and resources, to say nothing of the cultural values, differ from one country to another, leading to different decisions in similar situations.

It is particularly important to recognise that every energy source has its cost, not only in economic terms but in terms of safety and effects on the environment. There is no perfectly cheap, safe and harmless energy source. It is therefore highly misleading to call for perfect safety and then to argue that because a particular source is not perfectly safe, therefore it must be rejected. Although frequently heard, such demands are dangerous and irresponsible.

There is also the possibility of some new energy source that may transform the whole situation and solve all our problems. This is extremely unlikely, at least on the short term, and the deployment of a new energy source takes many decades before it begins to make a substantial contribution to world energy needs. There have been no fundamentally new energy sources since the discovery of nuclear fission about sixty years ago, and this is still far from being fully used. Nuclear fusion is still to be demonstrated as a viable option. Thus the only responsible way to tackle energy problems is in terms of known energy sources.

Finally, who is to make the decisions? In most cases it is the responsibility of Governments to formulate and implement an energy policy. This is not without dangers, as politicians are often concerned more with the next election than with the next generation. It takes five or ten years to build a power station, so the timescale of energy policies is much greater than the interval between elections in many countries. There is thus a serious danger that Governments will favour short-term solutions with the lowest cost, ignoring the harmful effects that may not appear for many decades. Governments are also very sensitive to the demands of pressure groups that frequently seek by false and emotional propaganda to influence how people think and how they vote. This may force a Government to take decisions that it knows to be wrong in order to placate public opinion.

This tension between short-term and long-term solutions to problems is quite general. It is natural for politicians to seek solutions to the immediate problems and ignore long-term ones. They do not want to spend money on projects that will reap benefits only in the next generation; they want to gain votes by doing things with immediate results. This is a serious defect in our democratic system, and it would be wise to form an all-party body with the responsibility to look further into the future. The perspective of this book is certainly long-term, though short-term effects are also mentioned.

Some energy problems such a pollution are international, because emission of toxic gases in one country can cause damage in another. The economic effects of energy decisions are often felt world-wide. Such problems have to be tackled on an international level such as that of the United Nations.

If all the decisions are taken at the Governmental or super-Governmental level, then what is left for us to do? In the first place we can play our part in a democratic society and urge that responsible energy polices be followed. We can see that we are not misled by pressure groups and do what we can to reduce their influence. Secondly, we can examine our own life-styles and see if we can reduce our energy consumption and avoid polluting our surroundings.

Scientists and other qualified people have a special responsibility for, depending on their expertise, they have a greater insight into scientific matters. University scientists are particularly well-placed to contribute to the public debate on energy matters, as they are employed neither by the energy industry nor by a political organisation. It must however be admitted that most of the contributions of scientists to the energy debate are neither welcomed nor appreciated, thus allowing errors to flourish unchecked.

The first step in a study of energy problems is to estimate the energy need, and the second is to see how this need can be met. The need depends on the number of people and the energy that each needs. We therefore begin by studying in Chap. 1 the population of the world and the way it is expected to grow in the future, and also the energy consumption. Both the rate of change of population and the energy consumption vary greatly from one country to another, and these differences are a vital part of the energy crisis.

Having established the need, the next step it to see how it can be met. There are many sources of energy that have been used throughout history.

Until the present century, the most widely-used sources were wood, and then coal, followed by oil and natural gas from the middle of the last century. These three traditional energy sources are discussed in Chap. 1. All these energy sources consume material that cannot easily be replenished, in the case of wood, or cannot be replenished at all, in the cases of coal, oil and natural gas.

The information we have on existing and projected energy supplies from these sources indicates that they will increasingly be unable to satisfy our energy needs. Unless new energy sources are developed, there is the likelihood of a serious energy shortage in the not-so-distant future. Since it takes a long time to develop new energy sources, action must be taken now. This is the basis of the energy crisis.

A possible solution is to reduce energy demand by conservation measures, but this will not make available sufficient energy to meet world-wide needs. It is therefore a matter of urgency to look at all possible energy sources to see if they can provide the energy we need. A very attractive option is to use the renewable energy sources such as hydroelectric, wind, wave, solar and geothermal that are practically inexhaustible, and these are considered in Chap. 2.

For different reasons, these renewable energy sources seem unlikely to be able to supply all our future energy needs. Another energy source is the nucleus of the atom, in particular the fission of the nuclei of some heavy elements, and this is discussed in Chap. 3. Another possibility, not yet realised but of great potential importance for the future, is nuclear fusion, and this is also discussed in Chap. 3. The safety of the fission reactors is discussed in Chap. 4.

All these possible energy sources must be examined to see what contribution they can make to our energy needs. To do this as objectively as possible, they must be evaluated according to their capacities, reliabilities, costs, safeties and effects on the environment, and this is done in Chaps. 5 and 6. There are many other threats to the environment, and these are also considered in Chap. 6.

Faced with the energy crisis and the threats to the environment it is desirable to outline our future energy policy. Should we abandon nuclear power and develop the renewable sources? Should we replace our polluting fossil fuels power stations with nuclear power stations? Or is there perhaps some intermediate policy? These questions are discussed in Chap. 7.

It is not sufficient, or even possible, to decide in an objective way what is the best solution to the energy crisis. The very criteria that are used depend on decisions that may well vary from one country to another, such as how much we are willing to pay for safety and to reduce the harm to the environment. There are many important issues where the scientific evidence is far from conclusive. Furthermore, even when the energy policy is agreed, there are still the political problems of putting it into operation. These are discussed in Chap. 8.

More fundamentally, all the problems related to our energy needs are moral problems, and the way we tackle them depends on our beliefs concerning the purpose of life. Are we free to plunder the earth to satisfy, not only our needs, but our pleasures and amusements, when poor people are starving? Should we care whether the pollution from our factories kills fish and trees in other countries? Do we have the responsibility not to bequeath to future generations a poisoned and exhausted earth? Our answers to such questions depends on our beliefs, and are discussed in Chap. 9.

1

THE ENERGY CRISIS

1.1 The Need for Energy

We need energy to cook our food, to light and heat our homes, and to drive our transport. We cook our food by burning wood, coal, gas, oil, or by electricity. We light and heat our homes with the same energy sources. We need petrol and oil for our cars, and coal or oil or electricity for our trains, ships and airplanes. The factories that make our clothes, our furniture, our refrigerators, our dishwashers, our packaged food all depend on energy.

Without energy all this would stop. Nothing could be made in factories. Cars and trains and ships and airplanes would no longer be possible. Our homes would be dark and cold, and we would have to eat raw food. All forms of radio and electronic communication would cease. Our lives would hardly be worth living; indeed we would not survive for very long.

The present population of the world is about 5.8 billion people, and this will rise to about six billion by the year 2000. On the average, it is doubling about every 35 years. People expect higher living standards, and as these depend on energy, the demand for energy is rising even more rapidly, increasing by 5% per year, equivalent to doubling about every fourteen years. The consumption of electricity, amounting to about a quarter of the total, is increasing even more rapidly, by 7 to 10% per year, equivalent to doubling every 10 to 7 years. Thus about half the increase in energy demand is due to the increase in population and half to the higher living standards, both doubling between 1953 and 1993.

All these are average figures; the reality is more complicated. As shown in Fig. 1.1, in some countries such as Europe and North America the growth of population is small, whereas in Africa and Asia it is much more rapid.

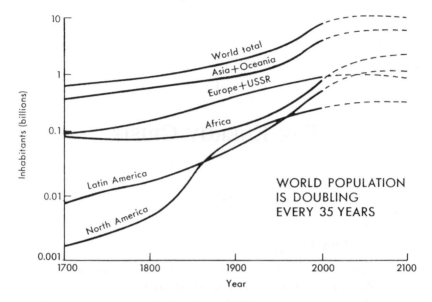

Fig. 1.1. World population growth.

It is obvious that these increases cannot continue for very long. If the present rate of increase continues, world population will reach the absurd level of 700 billion by 2150. It has to level off long before that, and the question is how it will happen, in a gradual way or by global catastrophes such as famine or wars initiated by the struggle for diminishing resources.

There are indeed some signs that the fertility rates and hence the rate of population increase is slowing down; the number of extra people in 1995 was about 81 million, compared with about 86 million in the late 1980's. Computer models indicate that world population may stabilise eventually at about 12 billion, over twice the present population. This is still very uncertain, but gives some indication of the magnitude of the problems facing us.

Detailed analyses of population growth show that the industrial development of a region brings about a demographic transition. At first the rate of growth is very high, and then it reaches a maximum and thereafter falls, so that eventually the population attains a fairly stable level. This has indeed happened in Europe and North America in the last century and the present century. At the present time the population in the developing countries is increasing rapidly, so if they can develop industrially in the same way, their rate of population growth will eventually fall, and the population will reach a stable level. Calculations indicate that if this happens the world population will reach 8 billion by 2020, 12 to 13 billion by 2150 and will eventually stabilise around 14–15 billion by 2150, about three times the present level. This population scenario depends on whether the developing countries do indeed achieve the necessary level of economic development. It also relies on sufficient similarity between the behaviour of people in the developing and developed countries. Furthermore, the most vital question is whether the developing countries can achieve this growth rate without intolerable devastation of the environment.

There are, however, some grounds for hoping that the developing countries may achieve a higher standard of living more rapidly than the developed countries. They can learn from the mistakes of the past and use more highly developed techniques. The developed countries had a high expenditure of energy in the early stages of their development, and this reached a maximum and then fell due to improved energy efficiency. The developing countries can learn from this and achieve their own higher standards without such a great energy expenditure.

There have been many attempts to estimate the maximum population that the earth can support in a sustainable way. This depends on the amount of arable land, which is severely limited, the crop yields, that have increased spectacularly during the last few decades, and the amount of water available, which is enormous but unevenly spread. Possible adverse factors are the depletion of soil fertility, the long-term effects of pesticides and the development of resistant strains of insect pests. Also very relevant is the expected life-style and the amount of waste. Most of these factors are practically impossible to quantify, so it is not surprising that estimates of the sustainable population range from about six billion to 150 billion. The consensus is somewhere in the range 8 to 12 billion, which is just about the same as the range of population estimates for the year 2050. This means that we may be near the absolute limit that the earth can sustain, and indeed may quite soon exceed it. If that were to happen, it is very likely that the earth would be so degraded by the efforts to support too many people that it would become incapable of supporting a much smaller number, perhaps around one billion. This degradation occurs when organic matter is burnt instead of being returned to the soil, when trees vital to prevent erosion are used as fuel, and when pesticides have to be continually increased in strength to control new strains of insect pests.

As will be shown, there is more than enough energy worldwide in many different forms to support a much larger world population at a higher level. The more critical question is whether this can be done without irretrievably damaging the earth. Some analysts think that the limits of the carrying capacity of the earth have already been reached. Despite the green revolution and the great increase in corn production during the 1980's, the annual growth per capita has fallen by 10% between 1984 and 1996. World fisheries have also reached the limit of sustainable harvests: the world catch has fallen by 9% since 1989.

There are large differences in living standards in different countries, so energy consumption is also very unevenly spread among the peoples of the world. The average energy consumption per person for the whole world is about 2 kW yr/yr (equivalent to about a gallon of oil per day per person), but about 70% of the world's population have to try to survive with much less than this. At the other extreme, people in the more developed countries use around 10 kW yr/yr. Since a reasonably high standard of living can be achieved with 4 kW yr/yr (the figure for Belgium), depending on the climate, it is apparent that there is much waste of energy. With improved efficiency of energy use, the figure could be reduced still further, but even then it will require an increase of world energy production by a factor of about three or four to bring the poorer countries up to the same level. Taking into account the expected increase of population by about a factor of three, this indicates that an increase of world energy production by about a factor of ten is required. This certainly cannot be achieved quickly, and is likely to be accompanied by intolerable pollution.

The total world energy consumption in 1995 was 13.4 TW, divided among the various sources as shown in Table 1.1.

Table 1.1. Energy Consumption in 1995 in TW (Holdren, 1996),[*] together with IAEA projections for 1980 and 2000 made in 1974 (Feld).[†] The three columns refer to the years 1980, 1995 and 2000.

Oil	4.1	4.6	5.2	Nuclear	0.6	0.8	9.4
Coal	2.4	3.0	4.1	Fuelwood		1.0	
Gas	2.0	2.6	3.3	Crop waste		0.4	
Hydro	0.6	0.8	1.6	Dung		0.2	

[*]J. Holdren, *Pugwash Newsletter*, January 96.
[†]B. T. Feld, *24th Pugwash Conference*, 249, 1974.

It is apparent from this Table that the projections made in 1974 were reasonably accurate, with the exception of nuclear, that has increased far less rapidly than expected.

Not only is the population of the world increasing, but so are the energy demands of each person. Each of us wants more and more energy. We want more clothes, more labour-saving devices. We want to go further away for our holidays. If these demands are met, then each year we use up more energy than we did the previous year.

The result is that the amount of energy being used in the world is increasing even faster than the population. Thus the energy consumption in the decade 1984–94 rose by 1.6% per year worldwide, 1.4% in Western Europe, 2.3% in Africa, 4.3% in China and 6.1% in India, and is estimated to reach 20 to 30 TW per yr by the year 2030. In 1996, the world energy use increased by 3%, including Thailand, South Korea and Pakistan by over 8%, South and Central America by 5%, Africa by 4.5%, Asian Pacific by 4.7% and the UK by 5.1%. Commercial energy use has increased tenfold this century. In 1996 the use of oil increased by 2.4%, natural gas by 4.5%, coal by 2.3%, nuclear by 2.3% and hydro by 1%. In 1986, the Soviet Union and other COMECON countries, including Eastern European countries and Cuba, were committed to increase their nuclear power production five or six times by the year 2000.

The actual consumption of energy does not however provide a true measure of world energy needs, because many people desperately need energy but do not get it. People in Europe, North America and Japan have a high standard of living and on the average use over ten times as much energy per person as people in poorer countries in Asia and Africa. In such countries many people have far less energy than they need, and so are cold and hungry. At the other extreme people like ourselves in the more developed countries use far more than we need to provide the basic necessities and even the reasonable luxuries of life. If we were to give all the people in the poorer countries as much energy as those in Europe, North America and Japan we would have to multiply the world energy production by a substantial factor. This would not only be economically very difficult, and would take many years, but would seriously deplete the remaining energy sources. In addition we would have to consider very seriously the effects on the environment.

If present trends continue, with a population growth of 1% per year in the rich countries and 2.5% in the poor ones, together with an energy growth rate of 5%, then by 2020 there will be 8.3 billion people in poor countries with an average energy use of 1.66 kW per capita and 1.9 billion in rich countries at an average energy use of 4.3 kW giving a total energy use of 80 TW, an increase by a factor of ten in 50 years. This unacceptable result could be avoided by increasing energy use in the poorer countries, reducing it in the rich countries and reducing population growth everywhere.

It should be remarked that most poor or developing countries are not at all homogeneous. There is usually a rich minority with lifestyles similar to those of people in the developed countries. The remainder of the population live in shanty towns around the big cities or in rural villages. Few of these people can afford electricity and so have to rely on wood for cooking and heating.

The energy needs of the rich and the poor are therefore entirely different. For the rich, more energy means a second or third car, more travel, more labour-saving devices and so on. For the poor, the availability of energy for the basic necessities is literally a matter of life or death. Most of the world's poor lack the energy for the basic necessities of life. For them, energy shortage means hunger instead of food, shacks instead of houses, cold instead of warmth, disease instead of health; in short, poverty instead of affluence. At the present time, this gap between rich and poor is widening.

This huge difference raises urgent questions. Is it right that such a gap should exist? What can we do about it? Can we ignore the poor and continue to live in luxury? Should we not reduce our energy consumption to make more available to the poor? Even if we were to do this, it would only be a small contribution to solving the problem. Whichever way we look at it, the world needs more energy.

It must be emphasised again that this is a long-term, global statement. Some countries desperately need energy now, while others have sufficient coal and oil to satisfy their needs for at least a few decades. It is similar to the connected problem of world food supplies. Although there is famine in some regions there is a global food surplus. Huge areas in Europe and North America are lying fallow, and even then there are meat and butter mountains and wine lakes of food for which there is no market. If more food were needed, it could easily be produced. The problem is the cost of production and transport, which is beyond the resources of the world's poor. In the short-term, the energy and food crisis is a political problem, not a scientific or technological one. For the longer term, it is not clear that even with maximum political will and all available scientific and technological resources that it is possible to find a solution.

Even if we adopt the relatively modest goal of increasing the average energy consumption to 3 kW yr/yr over the next century, we would have to greatly increase the world energy production, taking into account the increase of population in the same period. The disparity between energy consumption, and thus living standards, between the richer and the poorer countries is a measure of the ultimate energy need. Can this be achieved without destroying the environment?

Energy comes in many forms. We can get energy in the form of heat by burning wood and coal and oil. We can use this heat to boil water and make the steam drive a turbine to generate electricity. This is more convenient to use because it is clean and can easily be sent from one place to another. The turbine can also be driven by falling water, and this is hydroelectric power. We can also get energy from the wind, in some places from hot springs, and possibly from the tides and the ocean waves. More recently we have learned how to obtain energy from the nucleus of the atom.

Over the centuries, these energy sources were used one after the other, as shown in Fig. 1.2. The relative consumption of wood has fallen continually as the available supplies were exhausted. Coal has been important for several centuries, and is still a major source. Oil and natural gas are more convenient and their use has risen rapidly, taking over from coal. The available supplies are

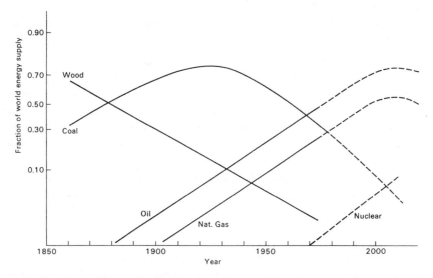

Fig. 1.2. The fraction of world energy supplied by different sources, showing how coal replaced wood, and then how oil and gas replaced coal. What will replace the oil and the gas when they eventually run out? (Hafele, 1981).

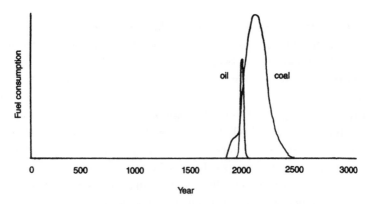

Fig. 1.3. Expected duration of fossil fuels, AD 0–3000. Oil and natural gas will last but a moment in man's history. *Source: Sir George Porter, President of the Royal Society.* (Perutz, 1989).

predicted to peak quite early in the next century, and then we are faced with the problem of replacing them by another source of energy. Viewed in the context of centuries, the fossil fuels can supply our energy needs for a rather short time: about fifty years for oil and gas, and perhaps three or four hundred years for coal, as shown in Fig. 1.3.

Which source of energy we use depends on a number of considerations such as availability, cost and reliability, and has to be decided for each region. We can get it more easily, cheaply and safely in some ways than in others. We therefore have to consider how much energy we can get from each source, and its cost, both direct and indirect. The direct costs are those of mining, extraction of the fuel from the ground, transport to the power station, construction and maintenance of the power station, and transmission of the energy to the consumer. The indirect costs include pollution and other effects on the environment, dangers to the health of the workers and to the whole population and the effects on the environment. To decide these questions we have to consider each energy source in turn.

1.2   Wood

Over the centuries, we have found increasingly concentrated and convenient sources of energy to satisfy our needs. From earliest times until the Middle Ages we relied mainly on natural fuels such as wood for heating, on animal and vegetable oils for lighting and on primitive windmills and waterwheels for grinding corn. The demand for wood soon exceeded the supply. Many forests in the Mediterranean region were cut down in ancient times and those of central Europe followed in the Middle Ages.

The effects of energy demands on the environment are as old as history. In ancient times the countries along the southern shores of the Mediterranean supported large populations. In Tunisia the oasis of El Djem, now surrounded by empty desert, has a building similar to the Colosseum in Rome, showing that in Roman times El Djem was the centre of a rich and well-populated agricultural region. Over the years the aqueducts that brought water from the Atlas mountains fell into decay, trees were cut down, the soil eroded and the once-fertile countryside reduced to a desert. Another example is provided by the forests of willows, birches and pines that once thrived in the Outer Hebrides until they were burnt down during the Viking invasions in the ninth and tenth centuries. They never recovered.

Wood is still used extensively as fuel in the developing countries, together with other organic material that would be better returned to the soil. It is a familiar sight in India to see walls covered with pats of dung drying in the sun. Wood is scarce, and women and children often have to spend much of their time going further and further afield to collect what they need. This not only denudes the soil of essential organic material, but wastes precious time that could be devoted to other tasks, particularly to education in the case of the children. Laws have been passed forbidding the cutting down of trees, but these are unenforceable. In half the countries of Africa, 70% of the energy used comes from wood. The poorer countries in Asia and Latin America also use much firewood, and the world consumption is still increasing at the rate of 1 to 2.5% per year.

The use of wood as fuel is strikingly different for the developed and the developing countries, as shown in Table 1.2.

There are proposals to establish plantations of fast-growing plants (collectively known as biomass) that could be harvested and burnt as fuel. This possibility is discussed in Sec. 2.8.

1.3 Coal

As soon as coal was found, it was used instead of wood. Increasing quantities were mined in many European countries, and provided most of the power behind the industrial revolution in the nineteenth century. Before the development of railways, it was inconvenient to transport coal in large

Table 1.2. The use of Wood as Fuel in 1974 (O. R. Davidson, 1981).

Area	Fuel Wood $\times$ $10^6 m^3$	Energy $\times$ 10^{15} J	Energy $\times$ 10^{15} J	%Fuel Wood
North America	17.6	170	77,763	0.2
Western Europe	32.3	312	45,161	0.7
E. Europe and Former USSR	99.7	964	54,267	1.8
Africa	268.3	2,594	1,848	58.4
Latin America	243.9	2,238	9,383	20.1
Far East	557.0	5,579	7,577	42.4
World	1,453.4	14,054	233,544	5.7

quantities from one place to another, and so the industries and people moved to the coalfields. The huge increase in the population of Europe in the nineteenth century, the development of manufacturing industries, railways and steamships, were all made possible by coal. There are still very large deposits of coal in many countries, enough for several hundred years at the present rate of consumption.

Coal will certainly remain a major source of energy for the foreseeable future. The technology is well understood and it is a familiar and accepted source of energy. There are however several disadvantages of coal that deserve serious consideration. Coal mining is dangerous, dirty and unpleasant, and increased coal production may mean more people working underground, and this must not be accepted lightly. In addition, as described in Chap. 6, coal burning on a large scale causes serious pollution.

The coal reserves in 1996 amounted to 1.03 trillion tonnes, corresponding to a lifetime of 224 years at the present rate of production.

1.4 Oil

In some places oil oozes out of the ground, and this has been known since ancient times. In began to be used in large quantities when oil wells were drilled. The first was in Pennsylvania in 1859 and thereafter production rose rapidly. Oil has a higher energy content and is much easier to extract from the ground than coal, and also to transport from the well to the consumer. It was therefore not necessary for the population and the industries to move to the oil wells. Instead, huge tankers carry the oil across the oceans of the world, from the Middle East and Alaska to Europe, the United States and Japan.

As the present century progressed, the use of oil rose rapidly, and in many industries it took the place of coal. At the same time it made possible a whole new range of industries such as plastics, drugs, paints, dyes and so on, as well as air and motor transport.

In the last few decades some huge new oilfields have been discovered, and in most cases the oil is easily pumped out of the ground. It is clean and abundant, and until recently it was very cheap as well. This is why oil has largely replaced coal as a source of energy in many countries. We have all become very dependent on oil, and this was brought home to us by the sharp rise in oil prices in 1973.

Not only is oil increasingly expensive, it is fast running out. Oil wells last about twenty or thirty years, and so we can estimate future oil production from the present rate of discovery of new oilfields. The chilling fact is that this rate of discovery is falling steadily as more and more of the earth is explored. No new supergiant fields like those in Alaska, which supply the bulk of the oil, have been found since 1975. It has been estimated that to maintain present oil production it is necessary to find new oilfields equivalent to that in the North Sea every two years. Geologists familiar with the poor rate of discovery during the last decade believe this to be impossible. This means that oil production cannot go on increasing for very long. In a few decades it is bound to level off and begin to fall. In the USA, for example, oil production peaked in the late 1970's and is now falling sharply. It is estimated that by the year 2000 most of the oil will be effectively exhausted, which means that it takes more energy to extract the oil from the ground than is given out when it is burnt. For the world as a whole, the reserves of oil will be exhausted by 2040 at the present rate of production, and by 2020 if production increases by 6% per year. Thus the best estimates indicate that world oil production will peak early in the next century and then fall rapidly. Unless something very drastic is done there will be a serious energy shortage. Since it takes a long time to develop new sources of energy we must tackle the crisis now.

Quite apart from the declining production, dependence on oil can be politically unwise. About 66% of the remaining proven oil reserves are in the Middle East and in 1989 the percentage of oil imports from that region were: 63% (Japan), 29% (Western Europe) and 11% (USA). This dependence has already led to the Gulf War. These political considerations are particularly critical because the Middle East oil reserves are expected to last about 100 to 150 years, whereas those in the rest of the world are likely to be exhausted in

Table 1.3. Proven Oil Reserves and Expected Duration.

Region	Quantity (Mil. Tonnes)	Share (%)	Duration (years)
Asia-Pacific	3,156	2.3	17
West Europe	2,635	1.9	14
Middle East	92,335	66.0	115
Africa	8,337	5.9	29
Americas	21,983	15.7	25
China	3,360	2.4	24
USSR	8,176	5.8	14
World	140,220	100	45

A forecast of world oil production during the next century is given in Fig. 1.4.

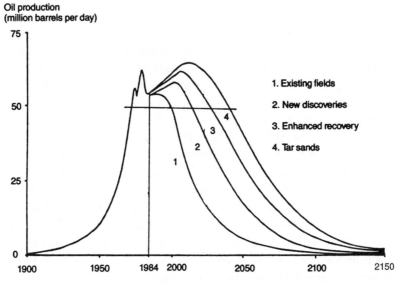

Fig. 1.4. Forecast of world oil procuction by Sir Peter Baxendell, managing director of Royal Dutch Shell. The curves indicate ratio of production in millions of barrels per day: (1) from existing fields; (2) from new fields likely to be discovered; (3) by recovery from existing fields and new fields of oil that require expensive new techniques which are uneconomical at present prices; (4) recovery of oil that is absorbed in tar sands and is uneconomical to extract using present processes. Alberta in Canada has large deposits. The horizontal line shows 1984 daily consumption. *Source: Peter Baxendell, "Enhancing Oil Recovery-Making the Most of What We've Got," Transactions of Mining and Metallurgy 94A* (April 1985) (Perutz, 1989).

15–30 years, as shown in Table 1.3. Thus the dependence on the Middle East oil must steadily increase.

A forecast of world oil production during the next century is given in Fig. 1.4.

1.5 Natural Gas

Natural gas is often found associated with oil, and also independently. It can be used for heating and lighting, but in the early days there was no practicable way of transporting it to where it could be used, and so it was often wasted. Subsequently, gas was made from coal and used for street lighting and for home heating and lighting. It has now been replaced for lighting by electricity, which is cheaper and more convenient, but it is still used for heating.

If it is available in large quantities, gas can also be burnt in power stations to produce electricity. The large gas fields in the North Sea have made natural gas the cheapest form of energy in Britain at the present time, and hence the 'dash for gas'. All the new power stations except Sizewell use gas as fuel. This is cheap and convenient while it lasts, but while the Government Department of Trade and Industry estimates a lifetime of some fifty years, other estimates are much less, predicting a decline by the year 2002.

The contribution of gas to world energy consumption has risen rapidly from less than 10EJ in 1963 to about 79EJ in 1993, accounting for about 21% of the world total. Like oil, it is easy to extract from the ground and can be piped over large distances. It can also be liquified and transported in refrigerated tankers by ship, road or rail. The proven reserves of gas are similar to those of oil, and the rate of consumption is only a little over one-half that of oil. As its consumption is rising rapidly, gas is unlikely to last longer than oil. It is being widely used not only for heating buildings and generating electricity, but in a wide range of chemical industries. Like oil, it is so useful that it is most unwise to exhaust our supplies by burning it.

A large gas field in Siberia is now supplying Western Europe with gas through a 5000-mile high pressure pipeline. In 1991, this supplied 20% of West European gas, including Finland (100% of total domestic supply), Austria (76%), Germany (34%), France (31%) and Italy (29%). In other countries it is more economical to liquify the gas for transport.

In the period 1992–1996 the total gas consumption in Britain increased by 45%. The increased dependence on gas is likely to result in price increases. The Energy Advisory Panel estimates that the price of 17.3 p/therm in 1990 is likely to rise to between 22.0 and 28.8 by 2005 and 25 to 37 p/therm by 2020.

The estimated natural gas reserves in 1996 amounted to 1412 trillion cubic metres, corresponding to a lifetime of 62.2 years. Of these reserves, 40.4% is in the FSU and 32.5% in the Middle East. For Britain, the estimated lifetime is 8.3 years.

In many respects, gas is now the cheapest, safest and most convenient energy source, but its lifetime is severely limited.

1.6 The Energy Crisis

At present most of the world's energy is supplied by coal, oil, natural gas and hydropower. Coal production could certainly be increased, but at great cost in health and damage to the environment. Oil and its associated natural gas will remain a major source for many years, but in a few decades will be insufficient. Hydropower is very useful, but because of the limit to the number of suitable rivers it can never supply more than around 10% of world energy needs. It is considered in the next chapter together with other renewable energy sources.

The seriousness of the impending world energy shortage is shown by estimates made in 1986, 1994 and 1996 of the proven reserves of coal, oil and gas for the United Kingdom and for the whole world. When these are divided by the present rates of production they give the number of years' supply listed in Table 1.4.

Table 1.4. Number of years' supply of fuel, estimated in 1986, 1994 and 1996.

Fuel	United Kingdom	World		
	1986	1986	1994	1996
Coal	114	125	197	–
Oil	14	34	40	43
Gas	16	58	56	65

These figures refer to proven reserves, and so will be increased if further reserves are found. Indeed, there is a definite tendency for these estimates to increase with time, indicating that such estimates are often pessimistic. On the other hand, they are computed using present rates of production, and so will fall if production rises due to increasing demand. If production rises to levels that significantly improve the situation of the world's poor, the reserves will be exhausted quite soon.

The basic fact underlying the energy crisis is that world energy demands are rising rapidly and yet the production of oil, our main source of energy, will soon start to fall. There will be a widening gap between energy supply and demand. In the next few years the world will face an increasing shortage of energy, and this will have the most serious consequences if a new source of energy is not found. Not only will the price of oil rise, but the scramble for the remaining oil supplies will be a potent source of international tension. Already the Middle East is a highly unstable area, mainly because it is known to hold more than half the world's remaining oil reserves. The impact of the energy shortage is very different from country to country. Some are well-endowed with natural resources, while others are already seriously short of the energy they need.

The gap between world energy needs and resources is so large that it certainly cannot be filled by any one source alone. It is thus not a question whether this or that source will solve the problem. We need energy from all practicable sources, and we must do all we can to develop new sources. The questions to be faced are: which sources are likely to be economical, which can provide the large amounts we need, and which can do no more than provide useful supplementary amounts for particular purposes.

Even if we were able to satisfy all our energy needs by greatly increasing world energy production there would still be serious problems, in particular the effects on the environment. It is obvious that world population and energy demand cannot go on increasing for ever. While we are rightly seeking to increase world energy production, we must also see how the demand itself can be moderated. There is no doubt that we waste a lot of energy, especially in the developed countries. We keep our homes too warm in winter, and use air conditioning too readily in summer. Many factories are using old and inefficient machinery, which wastes energy.

If we cannot produce the energy we need, can we solve the problem by reducing the demand by energy conservation? This will be considered in the next section.

1.7 Energy Conservation

At present we all use energy very wastefully. Most of the heat in our houses escapes through the walls and the windows, and industrial processes, developed in times of cheap oil, are far less efficient than they should be. Huge amounts of energy are spent on pleasure in some parts of the world while elsewhere people lack the energy for the bare necessities of life.

Considerable energy savings can be achieved at once simply by using less. Do we really need to have our homes at such high temperatures in winter, and to cool them so drastically in summer? By altering the settings on the thermostat we can maintain a perfectly acceptable temperature with much less fuel. Do we need to take the car for a short journey, instead of walking? Could we not take a bus or train for a longer journey, instead of the car? Is the journey really necessary? By moderating our lifestyle in these and many other ways, we can immediately save energy.

In addition, by installing thermostats, insulating loft floors, lagging pipes and using double glazing we can make further savings. Such savings can come only gradually because the insulating material itself costs energy to manufacture, so it is some time before we recover the energy spent. While it is expensive to improve the thermal properties of old buildings, new ones can be designed to be energy-saving at much less cost. We may therefore expect buildings of all types to become gradually more energy-saving over the years, as old ones are replaced by new. The potential savings are shown by the 40% increase in energy efficiency in the USA in the seventeen years following the 1973 oil price shock. Hafele considers that the cost of energy-savings measures sets a limit of about 43% on the energy saving that is possible in industrialised countries.

The motivation behind increased efficiency of energy use is extremely important, because the aim is to reduce energy use, not just to increase profits. The present rate of increase of energy consumption in many countries is far too high, and cannot continue indefinitely. The style of life that has been unquestioningly accepted in times of cheap energy must be critically examined, so that the limited energy supplies of the world can be more justly shared.

A difficulty about improving energy efficiency is that it generally leads to a reduction in price. This is good if means that poor people are now able to afford the energy they need, but it may also mean that people who already have enough for their needs will use even more on luxuries, and this always leads to increased energy consumption. This problem could perhaps be tackled by a system of differential tariffs, increasing the price with the level of consumption, so that the poor can afford what they need and the rich are discouraged from wasting it.

We must certainly conserve all the energy we can, and use what we need as efficiently as possible. The amount of energy saved, however, will inevitably be much smaller than the greatly increased demand. Energy conservation is highly desirable, but it is not enough to solve our energy problems on its own.

Even with the maximum energy conservation the world energy demand will continue to rise, though not quite so sharply as before. Without energy conservation, there is no hope of raising the standard of living of the world's poor to a reasonable level. Furthermore, it is necessary to reduce pollution by replacing power stations relying on fossil fuels. Since these now provide the major portion of our energy, there is an urgent need to look for other non-polluting energy sources.

References

Chesnais, J.-C., *The Demographic Transition*, Oxford, 1992.

Cohen, J. E., *How Many People can the Earth Support?* Norton, 1997.

Davidson, O. R., The Utilisation of Alternative Energy Sources, *Proceedings of the 31st Pugwash Conference*, 1981, p. 133.

Fremlin, J. H., *Power Production: What are the Risks?* Oxford University Press, 1987.

Hafele, W. (Ed), *Energy in a Finite World: A Global Systems Analysis*, Ballinger Publishing Co., 1981.

Mabro, R., *World Energy: Issues and Problems*, Oxford, 1980.

Meadows, D. H., Meadows, D. L., Randers, J. and Behrens, J., *The Limits to Growth*, Pontomac Associates Book, 1972.

Perutz, M., *Is Science Necessary?* Oxford, 1989.

Pindyck, R. S., *The Structure of World Energy Demand*, MIT Press, 1979.

Ramage, J., *Energy: A Guidebook*, Oxford University Press, 1997.

The Carrying Capacity Network and Briefing Book, Washington DC.

Wilson, C. L., Energy: Global Prospects 1985-2000, *Report of the Workshop on Alternative Energy Strategies*, McGraw-Hill, 1977.

2

THE RENEWABLE ENERGY SOURCES

2.1 Introduction

The energy sources considered in the last chapter all consume material that cannot be replaced except on a more or less long timescale. The forests of the Mediterranean and of central Europe were destroyed and cannot be replaced. Relatively small numbers of trees can indeed be replanted, but this is not an efficient way of providing new energy. If we set out to grow material to burn we would use other plants, and this is known as biomass and is considered below. Coal, oil and natural gas are irreplaceable natural resources. The amount available to us is finite, and once it is burnt, it is gone forever.

This suggests that if we want to solve the energy crisis we would do well to see if there are other energy sources that are renewable, that are not used up and so are inexhaustible. There are many such sources, including hydroelectric, wind, wave, solar, ocean and biomass, and they are considered in this chapter.

All of these derive their energy indirectly from the sun, and so are renewable only on the timescale of the sun's life. This is measured in billions of years, so large compared with our lives that it is essentially without limit. In the end, the sun will die, but that is not a problem to cause us any concern.

Geothermal energy comes from the decay of radioactive material in the earth, and so it is being gradually used up. However the timescale of depletion is so large that it is appropriate to include it among the renewables.

The term 'renewable' is sometimes used to include hydroelectric power. Although it is included in this chapter because it is indeed renewable, it is desirable to distinguish between it and the other renewable sources because it is a well-established source that contributes to world energy production at a level approaching 10%. This places it in a special category, between coal, oil and nuclear with much larger contributions on the one hand, and the remaining renewables with much lower contributions on the other.

Before considering the renewable energy sources in detail, it is useful to compare the contributions to world energy consumption from the various energy sources, and this is done in the next section.

2.2   World Energy Consumption

The contributions of the main energy sources to world energy consumption is shown in Fig. 2.1 for 1985 and 1990, together with a projection to 2000. This shows immediately that the renewable energy sources, apart from hydroelectric, make a very small contribution both now and in the foreseeable future. This is an inevitable consequence of the inherent limitations of the

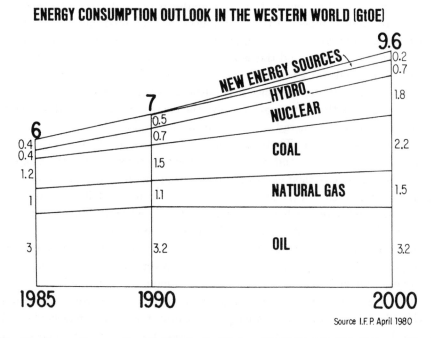

Fig. 2.1. Energy Consumption Outlook in the Western World (GtOE) (World Energy Needs and Resources).

renewables that will be described in more detail in the following sections. It is important to distinguish between hydroelectric power, which is substantial and well-established, and the other renewables, which are not.

The individual contributions of the renewables in 1990 were modern biomass 0.121, solar 0.012, wind 0.001, and geothermal 0.012 GtOE.

Not included in Fig. 2.1 are what are called the 'traditional' renewable energy sources such as wood, straw and dung. These amounted to about 0.9 GtOE in 1990. This is quite substantial, but it is certainly undesirable to try to increase the consumption of straw and dung. Indeed a decrease would be preferable, as this would mean that valuable organic material is returned to the soil instead of being burnt. In this context wood means growing trees that are subsequently cut down and used for fuel. It is only useful to consider increasing this if the trees can be grown on land that is unsuitable for agriculture.

The extrapolation of world energy consumption beyond 2000 is inevitably somewhat uncertain, but a reasonable attempt is shown in Fig. 2.2. This gives the percentage contributions of the various energy sources up to 2030. It is

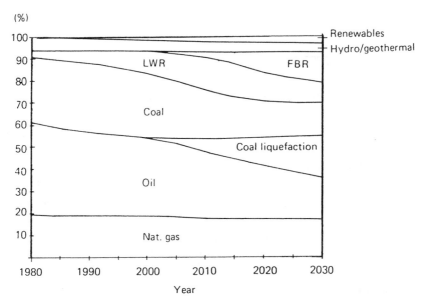

Fig. 2.2. Shares of primary energy sources in the global energy balance (World Energy Needs and Resources).

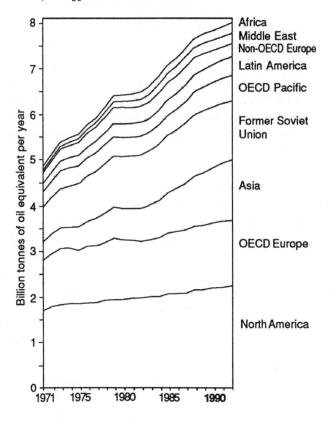

Fig. 2.3. World Primary Energy Consumption (International Energy Agency).

notable that natural gas remains about the same, while oil is decreasing. Coal, together with coal liquefaction, remains about the same. Hydroelectric power remains constant, and the renewables show some small growth. The contribution of nuclear power is rising quite rapidly, firstly due to the light water reactors, and then after 2000 from the fast breeder reactors. For the reasons to be discussed later, it now seems likely that the fast breeder reactors are unlikely to make an important contribution until well into the next century. Energy consumption increases at different rates in different parts of the world, as shown in Fig. 2.3.

These trends are seen clearly in the changing energy patterns for the European Community given in Table 2.1. This shows that during the period

1973–1989 the use of oil fell substantially and that of solid fuels and the renewables remained about constant. That of natural gas almost doubled and nuclear increased still more. For electricity production, nuclear power has now overtaken coal, and accounts for 35% of the total.

Table 2.1. Energy supplies to the European
Community in the period 1973–1989.

Fuel	1973	1979	1986	1989
Oil	646	622	505	509
Solid Fuel	232	235	230	229
Natural Gas	116	174	187	199
Nuclear	20	34	132	158
Hydro, Renewables	16	27	19	17

The patterns of energy use and their variations with time vary from country to country depending on their natural resources and their stage of development. Some instructive comparisons of energy use in Britain, Switzerland, India and the United States are given in the book by Ramage (1997).

It is also useful to compare the present world energy consumption of about 13 TWy/y (1994) with the maximum possible production from the renewable sources and with the remaining recoverable resources of non-renewable fuels. Table 2.2 shows that the plausibly harnessable renewable energy production is indeed able to provide the world's energy needs many times over. For comparison, the estimated remaining recoverable resources of fuels are petroleum 500, natural gas 500, heavy oils 500, coal 5000, oil shale 30,000, uranium (LWR) 3000, and uranium (FBR) 3,000,000 TWy. If fusion reactors are built, their potential contributions are from d-t fusion 140,000,000 and from d-d fusion 250,000,000,000 TW/y. It must be emphasised that these are estimates, and they only refer to what is physically possible, irrespective of all other considerations. There are of course many serious limitations that drastically limit what can actually be achieved, as will be discussed in detail.

Already established for many decades in the big league of power generators, hydroelectric is by far the most important of the renewables and so is considered first.

2.3 Hydroelectric Power

Electricity can also be generated by rivers in mountainous countries. To ensure a continuous supply of water, the rivers are dammed to form large lakes, and the water allowed to flow steadily through the turbines. Hydroelectric power stations provide much of the energy of countries like Switzerland and Norway that have many fast-flowing rivers. Most of the suitable rivers in the developed countries have already been used, but there are still many opportunities for the further development of hydroelectric power in developing countries, especially in South America, Africa and Asia. Many of their rivers are in regions remote from centres of population and industry, so the electricity generated has to be sent over large distances. This is not only expensive but is vulnerable to the actions of guerillas and saboteurs who blow up the pylons. This has frequently happened for the power generated by the Calboro Bassa dam on the Zambezi river.

Table 2.2. Plausibly Harnessable Renewable Energy Flows, (TWy/y at specified conversion efficiencies).

solar electric, 1% of land, 20% eff	50 (elec)
biomass, 10% of land, 1% eff	25 (chem)
ocean thrml, 2% of absorption, 2% eff	9 (elec)
hydropower, all practical sites	2 (elec)
wind power, windiest 3% of land area	1 (elec)
waves, ocean currents, tides, and geothermal energy all less than 1 TWy/y	
increased energy efficiency (2050)	10–40 (thrm)

Sources: Reports from World Bank, British Petroleum, UN, US Dept of Energy, US EPA, plus author estimates.
Total C stocks: atmospheric CO_2 = 760 GtC, biota = 700 GtC (J. Holdren, 1996).

Even if all the possible rivers are developed it has been estimated that hydroelectric power can never supply more than about 10% of the world's energy needs. Thus although it is very important, and deserves to be further developed in suitable places, hydroelectric power is not the way to solve the energy crisis.

Unlike coal and oil, hydroelectric power has the advantage that it is inexhaustible; it does not use up the resources of the earth. It is, however, very destructive of the environment.

At present hydroelectric power stations generate about 20EJ per year. In 1987 about 17% of electricity in the industrialised countries and 31% in the developing countries were generated hydroelectrically. This can certainly be increased, particularly in the developing countries, but is unlikely to exceed 60EJ per year. An estimate of the total capacity is three billion kW, of which about 9% is developed. Hydropower is relatively cheap because it has no fuel costs and rather small operating and maintenance costs.

In a few places there are areas of land below sea level that could be used to generate power hydroelectrically from sea water flowing into them. It has been estimated that 4,000 MW could be generated in this way in the Quattara depression in North Africa.

It is possible to store energy on a large scale by using two nearby lakes at different altitudes, as in Dinorwig in Wales.

2.4 Solar Power

The sun pours energy on the earth at the rate of 88,000 TW/yr, enough to satisfy our needs many thousand times. If we could find a practicable way of using it we would solve the energy crisis. Its heat stirs up the atmosphere and causes winds and waves, and these are a source of energy. We can also use the sun's energy directly, as solar energy.

On the average, the sun gives us about 200 watts per square metre at the earth's surface, with maximum values of up to 1000 watts per square metre in desert areas. Conversion losses reduce this by a factor of about four so that the average irradiation is to about enough to light one ordinary light bulb for each square metre. There is nothing we can do to increase this. We can easily work out the area of a collector required to satisfy the energy needs of an average family. Taking into account the efficiency of the equipment used, this has been estimated by Professor Hoyle to be about 1200 square metres. Thus a solar collector about the size of the Jodrell Bank radio-telescope would be required for each group of four families. To produce the same amount of electricity as a modern power station would require collectors covering about 50 to 100 sq. km., and to maintain efficiency all this has to be kept clean.

Solar energy can be harnessed in two ways, directly and by conversion into electricity. The simplest way to make direct use of the sun's heat is to put a solar panel on the roof of a house and let the sun's rays heat directly the water running through it. Aided by Government subsidies, many houses are now equipped with solar panels. It is necessary to have a normal boiler as well to boost the water temperature when required. Without a subsidy, such solar heating is marginally economical in Britain, but it can be more economical in sunnier countries.

Houses can now be designed so as to catch and retain as much as possible of the sunlight. Southern-facing windows and wall insulation can greatly reduce heating requirements. It is less easy to modify existing houses, but much can be done by loft insulation and double glazing. Such work itself costs energy, so it will be considerably later before this energy is repaid.

Solar ovens, with a large mirror to focus the sun's rays, are useful for cooking meals in sunny countries. While this is perfectly practicable, it means that meals can be best cooked only at midday, and not in the evenings, which is the traditional time in many countries such as India.

It is also possible to build huge mirrors to focus the sun's rays on a boiler and to use the steam to generate electricity. Such a generator takes up a considerable area and requires servomechanisms to move the mirrors continuously so that the rays remain focussed while the sun moves. It works only during the day and when the sun is shining, so it is both intermittent and unreliable, and it is far too expensive to be a practicable source of a large amount of energy.

There are however specialised applications where solar electricity generators are extremely useful. These occur when a relatively small amount of electricity is needed in locations where it is impossible or extremely expensive to provide it in any other way. Examples are road signs in remote desert areas and artificial satellites. These can be powered by photocells that convert the solar energy directly into electricity; the solar photons are absorbed by a semiconductor and the electrons that are released form an electric current. The photocells that are used to convert the sun's rays into electricity give a rather small voltage, so to increase this to a useful level it is necessary to have thousands of them in series.

As an example of the value of photovoltaic systems, the Georgia Power Company found that a photovoltaic lighting system costing $3000 removed the need for an extension of the electricity grid that would have cost $35,000. In the USA, several solar plants with a total capacity of 275 MW have been built in the Mojave desert, and 300 MW in California. The cost of the electricity generated has fallen from 23 cents per kWhr initially to an estimated 10 cents for plants constructed in 1990. It has been estimated that a 40 sq. m. array of photocells operating at 12% efficiency would suffice to provide the needs of an average household. Thus the total energy needs of the USA could in principle

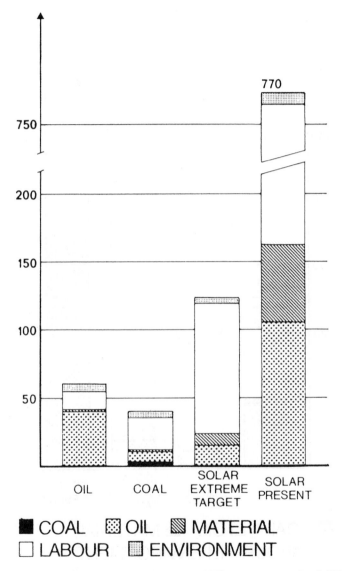

Fig. 2.4. Comparison of basic resources content a kWh at generation level (World Energy Needs and Resources).

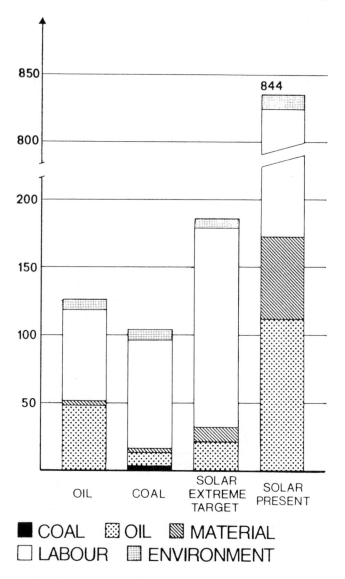

COST OF kWh AT DISTRIBUTION IN $10^{-3}\$(80)$

Fig. 2.5. Comparison of basic resources content a kWh at distribution level (World Energy Needs and Resources).

be supplied by a collector covering 34,000 sq. km. The cost of photovoltaic electricity is still rather high, but has fallen from $60 per kWhr in 1970 to around 20–30 cents in 1990, still five times that of other sources. The efficiency has improved from 15% in the mid-1970's to around 30% in 1990.

The costs of solar power at the generation level and at the distribution level are compared with those of coal and oil in Figs. 2.4 and 2.5. The solar costs are shown for the present state of development and for an extreme target development. It is apparent that the present solar cost in terms of oil is greater for an oil-fired plant; in other words it would be more economical, by a factor of more than two, to burn the oil itself instead of using it, along with other raw materials, to build a solar generator. Even with the most optimistic assumptions, solar is still more expensive than coal, and it would become competitive only if the oil is about 3.5 times its present cost, and coal over forty times. The mass production of photovoltaic panels is now driving the price down, and it is now estimated to be around 20 to 40 cents per kWh compared with 6 cents for conventional sources.

There is one other possibility that deserves mention, namely photochemical action. For example, light can split water into hydrogen and oxygen and these can be stored and then recombined when needed to produce heat. This is essentially what is done by plants: the oxygen is released to form the atmosphere which supports all higher forms of life and the hydrogen is used to convert atmospheric carbon dioxide to carbohydrates and other chemicals needed by plants. Unfortunately, however, there is no known process that can be used to photolyse water with sunlight on a large scale. It is however possible in principle and although some pioneering work has been done no real success has yet been achieved.

2.5 Wind Power

Windmills have been used to grind corn since ancient times, and they are still widely used to pump water up from wells on farms. In such applications it does not matter much that the wind is not always blowing because the water can be stored above ground until it is needed. It is easy to connect windmills to an electric generator to produce electricity. The new windmills are not like the old ones; they are large propeller blades mounted on high towers. About a thousand of these, spaced half a mile apart, are needed to equal the output of a coal power station. They are not reliable as a source since the wind does not always blow, and the cost of the electricity generated is greater than that from coal or nuclear power stations. Large arrays of windmills, called wind farms, have been built in the USA and in many other countries; at present this is only possible because they are subsidised. These take up large areas of land and are aesthetically objectionable, and so it has also been proposed that wind generators be situated on off-shore sites.

Wind energy has always been used for ship propulsion and continues to be important for fishing boats and yachts. Large ships require fossil fuels, although even for them some energy may be saved by adding sails for use when the wind is favourable.

Apart from hydroelectric energy, wind is the most promising of the renewable energy sources. It is however not easy to estimate the cost of wind power because so few windmills have been operated long enough for reliable statistics to be obtained. The technology is relatively new, and there are many designs that are being studied. Modern windmills are usually two-blade propeller units mounted on high pylons, and such units can power 200 kW electric generators. Experimental generators have produced as much as 2 MW at wind speeds of

Table 2.3. Wind energy in 1995 in MW (Wind directions, XV No 2 January 1996).

Country or region	Capacity at end of 1994	Estimated new capacity in 1995
USA	1722	150
Latin America	10	175
Germany	632	300
Denmark	539	75
Netherlands	162	40
United Kingdom	170	20
Spain	73	90
Sweden	40	15
Greece	36	10
Italy	22	10
Portugal	9	5
Ireland	8	10
Finland	4	3
Europe (Total)	1723	588
India	201	400
China	29	50
Africa (Total)	16	10
World (Total)	3738	1253

Table 2.4. Wind power in Europe in 1994 (*Nuclear Issues*, December 1996).

Country	Capacity MWe	Output GWh	Load factor %	Share of electricity %
UK	148.5	337.0	25.91	0.11
Sweden	45.0	72.0	18.26	0.05
Portugal	8.3	17.0	23.38	0.06
Netherlands	157.0	238.0	17.31	0.02
Italy	21.0	6.3	3.42	0.003
Ireland	6.5	18.0	31.61	0.11
Finland	5.0	7.0	15.98	0.02
Greece	26.9	37.4	15.87	0.10
France	3.4	9.0	30.22	0.002
Spain	74.9	175.2	26.70	0.11
Germany	643.0	1428.0	25.35	0.29
Denmark	532.0	1137.0	24.40	2.99
Belgium	5.2	8.6	18.88	0.01
Europe (13)	1671.7	3566.55	24.35	

12 m/s. The power obtainable from a windmill increases as the cube of the wind velocity, so a high average wind speed and good siting are essential. The power output of a windmill thus depends critically on the assumed average wind speed. Typical energy conversion efficiencies are about 20–40%, and at a wind speed of 7 m/s a windmill can intercept about 200 W/m^2.

Denmark is ideal for wind power but short of other energy sources, and it has been estimated that 10% of Denmark's electricity could be generated by wind by the end of the century. A large windmill near Aalborg has a power output of 630 kW and cost £500,000; smaller ones generating 30 kW cost £10,000. These costs are so high that the wind programme is only possible with Government subsidy. In 1991 Denmark had 360 MW installed capacity, The Netherlands 55, Germany 55, Spain 15 and the UK 10 MW. These figures have now increased substantially, as shown in Table 2.3. Nevertheless, the contribution of wind power to electricity generation is still very small, as shown in Table 2.4.

In Britain, a wind generator of 1 MW capacity having blades with a spread of 200 ft or more on a tower 150 ft high is planned. Between one and two thousand such generators would be needed to match the output of one large power station, and it would be able to do this only when the wind velocity is in the required range. Even larger wind generators are being built on the Orkney islands, but experience in many countries has shown such machines to be unreliable because the forces on the bearings are very high, and it may be beyond the capacity of the available materials to withstand them.

The largest concentration of wind turbines is in California, due to the tax incentives and the high prices paid by the power companies. At present there are 30 wind farms in the UK with a total of 505 turbines and a life expectancy of 25 years with a total capacity of 190 MW. The electricity they could generate in that time is equal to that produced by a coal power station in 3.5 months. Another 31 wind farms, with a total capacity of 300 MW are under construction or awaiting construction. Due to the variability of the wind, power is only produced for part of the time (see Fig. 2.6), and is difficult to integrate into the national grid. The load factor varies from place to place, but is usually about 20 to 30%. Taking this into account, wind power now contributes about 0.05% of the electricity used in the United Kingdom. In 1995, Sweden produced 0.17 TWh from 220 turbines with a total capacity of 70 MWe, with an average load factor of 16% at a cost of 30 to 40 ore per kWh. The cost of nuclear power in Sweden ranges from 13.3 to 21.5 ore per kWh.

An experimental 3 MW wind turbine at Growian in Germany was abandoned after less than two years' operation. It produced 80,000 kWh during this time, enough to meet the needs of 23 households at a cost of about a million pounds each. A wind farm has been proposed at Rookhope Common in County Durham comprising 40 turbines of 1.5 MW each, mounted on columns 295 ft high. Several environmental organisations have protested against this plan.

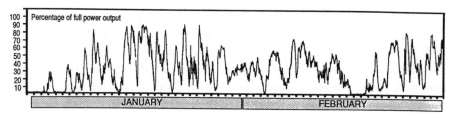

Fig. 2.6. The load factor for the combined output of five wind parks in Denmark with a total capacity of 28.2 MW for January and February 1990. The generators operate only with wind speeds in the range 4 to 25 m/s (*Nuclear Issues*, April 1990).

To generate the same amount of electricity as the Sizewell nuclear power station would require about a thousand wind turbines spread over 2000 sq km, which is about the size of East Suffolk. Each turbine would require a power line to connect it to the grid, and service roads for construction and maintenance. Permission for all this would be required in each case, and frequently this would encounter determined opposition.

Windmills inevitably occupy large areas of high ground and so are visible from large distances. In operation they tend to emit a humming or swishing noise. The noise level is quite low, and would hardly be noticed in a city. Wind farms are however often sited in quiet countryside where any noise is noticeable and some people find it unbearable. Having moved to a remote place for solitude, they soon want to move away, but cannot do so because their house is now unsaleable. Most proposals for new wind turbines encounter fierce local opposition, and many organisations concerned with the countryside, such as the Council for the Protection of Rural Wales, are strongly opposed. Mr. N. Caldwell, its Director, is quoted as saying that we are paying more for our electricity to manufacturers "so that they can cover our beautiful Welsh hills with turbines that only produce a trickle of energy". It has been argued that, apart from the visual and aural aspects, the turbines will give people headaches, ruin TV reception and kill thousands of birds; there is however no evidence for any of these effects. When a wind farm is being planned, the noise level at nearby houses is kept within 5 dB of the background noise without the wind turbines, and in very quiet places a limit of 35 to 40 dB may be required. Wind turbines do not operate for wind speeds below a certain minimum, and this wind itself raises the background level. It is often found that when the turbines are built, most people quite easily adapt to their presence, and recent surveys have shown a high degree of acceptability, and it is reasonable to expect that with improved design the noise levels will continue to decrease. It has been claimed that the noise level at 110 yards from a TW 600 generator is 43.5 decibels, quieter than a modern refrigerator.

Most of the commercial windmills in operation today were built with Government subsidies provided to stimulate research and development. Unfortunately electricity from windmills costs more than that from coal or nuclear power stations, and so as soon as the subsidy stops it is no longer economical to go on, and they are abandoned, or sold to optimistic Third World countries. This is now happening in California, site of some of the largest wind farms. It is difficult to obtain reliable estimates of the cost of wind energy, partly because the latest designs have obviously operated for only a fraction of their lifetimes (which are not known) and the costs depend sensitively on the rate of interest required on the capital invested. It is therefore likely that figures for comparative costs are more reliable than those for absolute costs. Thus it has been estimated that the cost of wind-generated electricity is about 50% more than that from conventional power stations (*Wind Directions*, January 1996). The costs of wind energy have fallen markedly in the last few years, from about 10 p/kWh to 4 p/kWh.

Wind power in operation poses little or no danger to the public, apart from the rare event of blade detachment. The wind turbines themselves are constructed in factories and this inevitably incurs the hazards of all manufacturing processes, together with the resultant atmospheric pollution. Several fatal accidents have occurred in the construction and maintenance of windmills. In 1991 two workers were killed in the Rodbyhaven wind farm in Denmark when their maintenance platforms struck one of the blades. The statistics are poor, but they indicate an accident rate of about two deaths per GWyr, which is comparable to coal power. Other hazards of wind power are lightning strikes, fire, storms and subsidence. It has been estimated that in Denmark about 6% of the turbines will be hit by lightning each year. The resulting damage to the blades is now minimised by incorporating conducting material in the blades so that the electrical surge is dispersed to the hub and thence to the ground.

Wind power is particularly useful on remote windswept islands and mountainous regions where the electricity demand is insufficient to justify a large power station. Electricity in such circumstances is usually provided by a diesel generator, and this can be retained as a backup when the wind fails. Such a system is operating on the Norwegian island of Froya. It consists of a 55 kW wind turbine combined with a 50 kW diesel generator. In the period from January to September 1995 the wind turbine operated for 66% of the time producing 78,000 kWh of energy, while the diesel contributed 39,400 kWh. There is a similar development on a small island in Guadaloupe, where a high plateau experiences winds of about 10 m/s for more than 80% of the time. A group of twenty turbines with a total capacity of 500 kW, together with a diesel plant as backup, now provides all 1600 inhabitants with their electricity needs. If a cyclone is forecast, they can be dismantled and laid flat on the ground until the danger has passed. A relatively small turbine producing 2.2 kW at a wind speed of 12 m/s provides light, power and water heating for a remote Youth Hostel in Scotland. In calmer weather, batteries provide a backup.

Offshore wind farms have the advantages that the winds are usually stronger and less turbulent and they do not take up agricultural land or cause any annoyance to people. On the other hand, they are less easy to build, even if the water is very shallow, are not so easy to maintain, may pose a danger to shipping and require an undersea cable to bring the power generated to the shore. To minimise transmission costs, the site chosen should not be too far from the shore. The first offshore wind project in 1990 was a 220 kW turbine sited 0.25 km from the shore in Sweden. This was followed in 1991 by a 4950 kW wind farm off the Dutch coast. The early experience was encouraging, though the first Swedish turbine was lost by fire. Some UK offshore wind farms are being planned, but large scale development is unlikely as they cost about 30% more than onshore wind farms.

It is often argued that wind power is non-polluting. This does not take into account that the windmills and associated equipment have to be made in factories, and the production processes are inevitably polluting. It has not been possible to obtain estimates of the pollution from this source. Wind farms are not only visually and aurally polluting, but the ground is broken up to make the massive concrete foundations and access roads, so that it is very costly to restore the site to its original condition after the windmills have been abandoned. It is also very difficult and time-consuming to obtain the necessary planning consent.

2.6 Tidal Power

The tides are a practicable source of energy in a few places such as the La Ranche estuary in France and the Severn estuary in Britain. The water is held back by a dam at high tide, and then allowed to run through turbines when the water level falls. A small 240 MW tidal power station with an average energy output of 65 MW has been operating for many years at La Ranche, but it is not economic. A detailed study has been made of the feasibility of using the Severn estuary, which is much larger and could produce up to 7 GW. It is estimated that the energy it would produce would cost about twice that of a conventional power station. An investment of about fifteen billion pounds spread over about ten years would be needed before any power is produced, and the environmental effects are likely to be severe. There is also the difficulty that the power is not always produced at the time of day when it is needed, since the tides follow the lunar cycle.

A more modest proposal to establish a tidal plant with an estimated capacity of 100 to 135 MW in the Dudden estuary in Cumbria ran into severe environmental opposition. According the protesters, the tidal barrage would ruin a "dramatic and wild landscape that is one of Britain's last wildernesses". This area of salt marshes, sand dunes, sand and mud is the habitat and staging post for wild fowl and migratory birds.

To be economic, a tidal power station must be built on a large scale, and the massive initial investment must be available at a low interest rate. The operating costs are high due to corrosion of the turbines by sea water and the growth of barnacles and seaweed. These factors have discouraged the building of further tidal power plants.

2.7   Wave and Ocean Thermal Power

The energy in ocean waves comes from the winds, and this in turn comes from the sun. The energy given to the ocean in this way has been estimated to be 2.7 TW per year. Many devices have been proposed to convert this vast amount of energy into a useable form, but so far none has operated commercially on a large scale. Inevitably the devices have to be very large and therefore very costly, and they are subjected to corrosion by sea water and to continuous buffeting by the waves. Thus a £3.5 M wave power generator Osprey 1 weighing 8000 tons sank in shallow water off the Scottish coast in a freak summer storm. It was designed to produce 2 MW of electricity, but never operated.

The energy recovery time from wave power generators is likely to be long, possibility in the region of twenty years, and the lifetime of the equipment may not be as great as this. There are in addition the costs of transporting the electricity generated to the shore, the cost of maintenance and the hazards to shipping.

Many detailed studies of wave generators have been made, including the use of scale models, but so far the results have been disappointing. Thus at present it seems unlikely that wave power will ever be economically practicable as a source of energy.

There is a considerable difference in temperature between the surface and the depths of the ocean and this can be used as a source of energy. In tropical regions the surface is at about 25°C and at about 1000 ft., the temperature is 5°C, and this temperature difference is sufficient to drive a turbine. The feasibility of this process was first demonstrated in the 1930's by the French scientist Georges Claude. More recently the idea has been developed by

Abraham Levi and a 50 MW plant has been constructed off Hawaii. A 1 MW prototype has been designed and is estimated to cost $40,000 per kW.

The difficulties of these devices are the corrosiveness of sea water, the low efficiency of energy conversion, the problem of anchoring the device in place and of transmitting the power generated to where it is needed.

2.8 Biomass

Large amounts of energy are obtained from burning wood and animal dung, and from the muscle power of animals, particularly in more primitive societies. There are few reliable estimates of the energy generated in this way. These methods are inefficient except on the level of domestic firewood, and as development proceeds they are replaced by more efficient sources of energy.

Such sources cannot solve the energy crisis, but it is possible to consider generating energy on a large scale from organic material, available either as a by-product or specially grown for the purpose. The total energy now obtained from biomass, including wood, is estimated to be 1.5 million TJ, but the efficiency is so low that the useful energy is less than 10% of this.

An example of the former is the use of the straw remaining after corn is harvested. If this were all burnt it could provide as much heat as 2% of Britain's oil imports. As straw is rather cumbersome to transport, it is best burnt on the farms, and many farmers have installed straw boilers. These can be used to heat farmhouses and animal buildings and to dry grain and hay. In France there is a project to build a plant to produce methanol from straw for use as tractor fuel. There are however disadvantages in using straw. The boilers are expensive and require automatic stoking machinery. Most farmers prefer to burn it in the fields, which fertilises the soil and kills weeds and insects. In Brazil the bagasse remaining after the sugar cane has been crushed is often burned in the factory to heat the water used to purify the sugar or in gas turbine systems to produce electricity.

Organic and inorganic waste material can be converted to synthetic fuels by several processes, with a net average yield of about one or two barrels of oil per ton of waste. At present however it is cheaper to burn the waste than to convert it into oil. At the present time solid wastes provide about 3% of the energy of the USA.

It is possible to grow crops with the sole purpose of burning them for energy. This is practicable but the efficiency of the process is very low. A high yield crop such as sugar cane has a productivity equivalent to a solar conversion efficiency of about 1%. Harvesting the biomass is expensive, so the efficiency of converting the biomass to fuel ranges from 20% to 50%, so that the overall efficiency for the conversion of biomass to synthetic fuels is around 0.4%. To meet the primary energy needs of the earth in this way would require good agricultural land covering 10% of the surface of the earth.

In Brazil, 62% of the fuel used in transport comes from alcohol extracted from sugar cane. This replaces 100,000 barrels of oil per day, and saves the country about $15B per year of petroleum output. This project has been criticised as a monoculture that has hindered the modernisation of rural economy and keeping the cane-cutters in semi-slavery.

While it is useful to generate energy from waste material and unwanted wood, too much valuable organic material is at present being burnt, especially in the poorer countries. This depletes the soil of its nutrients and eventually exhausts its fertility. In many countries the increasing demand for firewood is causing extensive deforestation, and replanting becomes almost impossible because young trees are torn up for firewood almost as soon as they are planted. This process can only be halted and reversed by the provision of alternative energy sources.

2.9 Geothermal Energy

The inside of the earth is hotter than the outside, and this becomes evident in regions where the earth's crust allows some of the heat to escape from volcanoes and hot springs. This water is sometimes hot enough for it to be used as a source of useful energy, although more often it is used for medicinal baths and as a tourist attraction. The amount of useful energy that can be obtained from a hot spring is quite small.

Geothermal energy is economical if large amounts of very hot water are readily available, or if it can be used directly for domestic heating. Relatively small amounts of electric power have been generated geothermally for many decades. In 1976 the total electrical capacity amounted to 1,325 MW.

On the average, the temperature of the earth's crust increases with depth by about 25°C per km. This heat is due to the radioactive materials in the earth. The amount of energy there is enormous, but it is costly to extract from the earth. This can be done in several ways, depending on the conditions. If the rock is dry and porous, two holes are drilled a convenient distance apart so that cold water can be pumped down one hole, driving hot water up the other. If the rock is wet, the hot water can be pumped up directly. After a time, the rock around the holes is cooled and is no longer useful as a heat source. Particularly in the case of dry rock, it takes a very long time for it to warm up again by conduction from the heat of the surrounding rocks. Thus the amount of heat that can be obtained from a particular hole is limited. It is extremely costly to bore such holes, and to ensure that they are the optimum distance apart. If they are too close, the amount of available energy is small, whereas if they are too far apart the water pumped down one hole cannot get through to the other. Many experiments have been made to assess the economic practicability of this type of geothermal power, and the results are disappointing.

A study of the feasibility of geothermal power in Cornwall was abandoned after an expenditure of £33 M. The energy obtained was estimated to cost four times as much as that from wind or solar. An application to build a geothermal power plant at Heber in California was refused because the Public Power Commission found that "the power delivered would be in the range 17.9 to 24.3 cents/kWh, compared with 11 cents/kWh for a coal fired plant and 16.6 cents/kWh for existing oil-fired generation".

The economic use of geothermal energy is thus limited to the rather few places where very hot water or steam escapes from the earth. The total amount of energy generated in this way is one or two thousand MW and so it is evident that it can make only a very small contribution to world energy needs.

The water from hot springs frequently contains a high concentration of dissolved salts that could pose a serious environmental problem. These salts must be removed from the water before it is released, and this adds to the cost of the process.

2.10 The Renewables

With the exception of hydropower, all these renewable energy sources are either costly or unreliable, or available only in particular places on a small scale, so none of them can supply energy on the scale we need.

The main disadvantage of wind and solar power is that they are inevitably unreliable. Sometimes the wind blows and sometimes it does not, and clouds often obscure the sun. This would not matter quite so much if we could store the energy, but unfortunately there is no practicable way of doing this economically on a large scale. We have to generate the power when we need it, even on quiet, cloudy winter days. So if we went in for solar or wind power on a large scale we would still need to have coal or oil or something else as a standby. And since wind and solar are rather more expensive than coal or oil, the only argument in their favour is the relative absence of pollution.

The basic reason for the high cost of the renewable energies is very simple: the sun's energy is spread very thinly over the earth's surface, so we have to go to much trouble to concentrate it. It is very hard to break even when it comes to cost. It is only when nature does the concentrating for us, as it does with the fossil fuels and with hydroelectric power that it becomes a practicable source of energy on a large scale.

Wind power is the most promising of the renewables, and many experimental wind generators have been built. It is important to continue this research and to develop new types of wind turbines.

The cost of wind power has fallen considerably in the last few years and it is certainly desirable to extend its application wherever practicable. The development of wind power on a large scale is likely to encounter economic and environmental difficulties. It is not possible to make a reliable estimate of the future contribution of wind power simply by multiplying existing figures for small-scale production. Thus it has been estimated that the US energy demand could be met by four million 50 kW wind turbines spread in a grid 0.5 km apart. Such statements do not take adequate account of the difficulties of implementing such a project and could be termed the multiplying-up fallacy.

The more optimistic hopes for wind power envisage that wind might provide about 10% of Europe's electricity needs by 2030. At the present level of power generation they are not in the same league as coal, oil and nuclear, and they cannot provide a solution for the world energy crisis.

The renewables are often called 'benign', but this is hardly deserved. In addition to the visual and aural pollution, they are relatively quite dangerous. This will be quantified later on when we compare the safety of various methods of power generation. The hazards associated with wind power are those normally associated with the manufacturing industries, together with those associated with the construction and operation of the windmills. Since large numbers of windmills are needed to produce a significant amount of energy, the cumulative hazards per unit electricity produced compares rather unfavourably with those of other energy sources, with the exception of coal. Similar remarks apply to solar power and to the other renewables.

There remains the possibility of nuclear power, and this is considered in the next chapter.

References

Blair, I. M., Jones, B. D., and Van Horn, A. J. (Eds), *Aspects of Energy Conversion*, Pergamon, Oxford, 1976.

Boyle, G. (Ed), *Renewable Energy; Power for a Sustainable Future*, Oxford, 1996.

Blunden, J. and Reddish, A., *Energy, Resources and Environment*, Hodder and Stoughton, London, 1991.

Hafele, W. (Ed), *Energy in a Finite World: A Global Systems Analysis*, Bellinger Publishing Co., 1981.

Miller, G. T., *Sustaining the Earth*, International Thomson Publishing, 1994.

Rooke, D., Fells, I. and Horlock, J. (Eds), *Royal Society Symposium*, London, 1995.

Wind Directions, Newsletter of the European and British Wind Energy Association.

Wind Energy: Power for a Sustainable Future, Policy Statement of the British Wind Energy Association, 1991.

World Energy Needs and Resources, Council for Science and Society Report, *Deciding about Energy Policy*, 1979.

3

NUCLEAR POWER

3.1 Introduction

There is another source of energy that we must consider, the nucleus of the atom. In 1939 it was discovered that the nuclei of some very heavy atoms like uranium 235 can break up into two nearly equal pieces, as shown in Fig. 3.1. This process is called fission, and the pieces are called fission fragments. They fly apart at high speed and this produces heat. Each fission also releases two or three neutrons, and these can trigger more fissions. This process continues very rapidly, with great release of energy. If the uranium is very concentrated the reaction takes place so quickly that there is a violent explosion. This is what happens in the atomic bomb.

It is also possible to dilute the uranium so that the energy is released gradually in a controllable way. This was first achieved by Fermi in 1942 when he built the first nuclear reactor. The energy of the fission fragments produces heat, and this can be used to boil water and drive a turbine to generate electricity just as in coal and oil power stations. The fission fragments are highly radioactive and so they must be prevented from escaping and harming the environment. The design and operation of such nuclear reactors is described in Sec. 3.2.

This is a new source of energy, and it uses a material, uranium, for which there is no other use, except on a very small scale to colour glass. It must therefore be assessed critically to see if it can supply our energy needs.

An important question is whether there is enough uranium at an economic price to fuel the nuclear reactors that will be needed in the next century. Uranium is very widespread, but the concentration of uranium varies from one

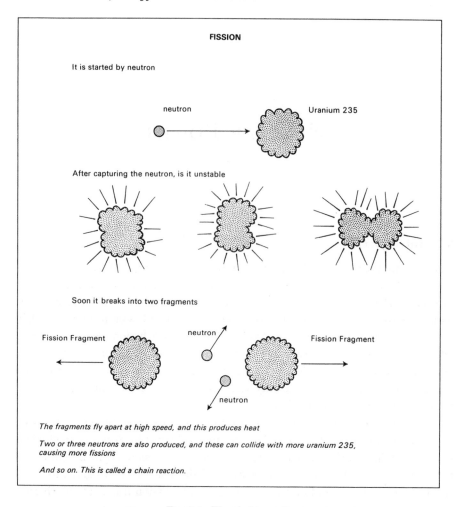

Fig. 3.1. The chain reaction.

deposit to another. The amount we can extract depends on the price we are prepared to pay for it, since the higher the price the more ores are economically workable. There is certainly enough uranium to last for several decades at the present rate of production, and as this is exhausted thorium can also be used.

If that were the whole story, the development of nuclear power would certainly ease the energy crisis, but would not solve it. In about a hundred years, when all the useable uranium and thorium are exhausted, we would once again face a severe shortage of energy.

Fortunately there is another way of burning uranium, known as breeding. Natural uranium consists of two types of uranium nuclei, and only one of these, uranium 235, is easily fissile. This is present in natural uranium in a very small proportion, less than 1%. The present nuclear reactors, called thermal reactors, burn only this 1% of the uranium. It has now been found possible to design a reactor, called a fast reactor, that is able to burn the remaining 99% of uranium 238 as well. Allowing for losses, this effectively multiplies the amount of fissile material available by a factor of about fifty. The uranium is converted into plutonium 239, which is also fissile and so can be used as a reactor fuel. Several large fast reactors have already been built, and they will be available to take over power production when the price of uranium rises to the level at which it is economical to do so. These fast reactors are also discussed in Sec. 3.2.

Nuclear reactors differ from all other energy sources because the energy comes from the nucleus of the atom. In the process many highly radioactive nuclei are produced, and these emit nuclear radiations that can damage living organisms. These nuclear radiations are described in Sec. 3.3.

Nuclear reactors are inherently safe, but if they are badly designed or operated in defiance of the safety instructions, disasters can occur. The most serious were those at Three Mile Island and at Chernobyl, and these are discussed in the next chapter.

Nuclear reactors produce highly radioactive fission fragments as a by-product, and the way they can be disposed of safely is described in Sec. 3.4. Nuclear power has several other applications, and some of these are listed in Sec. 3.5.

Using fast reactors, there will be enough fissile material available for the foreseeable future. It is no use thinking more than about fifty years ahead, because before then there may be new technological developments that completely transform the situation. One of these is the fusion reactor, and this is discussed in Sec. 3.6.

3.2 Nuclear Reactors

All matter is composed of atoms, and each atom has a tiny central nucleus surrounded by orbiting electrons, rather like planets going round the sun. Nuclei are exceedingly small, to an extent that defies the imagination. They are roughly spherical and have diameters of a few 10^{-13} cm, a unit known as the fermi, after Enrico Fermi, a pioneer of nuclear physics. To see what this size means, imagine a row of nuclei placed side by side in a row across one of the full stops on this page, like beads on a string. Now enlarge the full stop to the size of Europe. The nuclei would still be too small to see; there would be ten thousand of them per centimetre.

Nuclei are complicated structures composed of particles called protons and neutrons, collectively called nucleons, and tightly bound together by the nuclear forces. Each proton carries one unit of charge, whereas the neutrons are uncharged, so the number of protons Z is the nuclear charge. Each electron also carries a unit of charge and as the atom as a whole is neutral, the number of electrons orbiting the nucleus is also Z. If N is the number of neutrons, the total number of nucleons in the nucleus is N + Z = A, the atomic mass number. The neutrons and protons have very similar masses, and are nearly two thousand times as heavy as the electron, so nearly all the mass of the atom is concentrated in the nucleus.

The nucleus of hydrogen is a single proton and that of a heavy nucleus may comprise up to about eighty protons and a somewhat larger number of neutrons. The most stable nuclei, the most tightly-bound, are those of medium size like iron with about fifty or sixty nucleons. This means that energy is released if we divide a heavy nucleus into two more or less equal pieces (called fission fragments) or if we stick together some light nuclei to make a heavier one. The first process is called fission, and is the basic process occurring in so-called atomic bombs and in nuclear reactors. The second is fusion occurring in hydrogen bombs and in the future, it is hoped, in fusion reactors. The fission process occurs in some very heavy nuclei when they are struck by low energy neutrons. These are called fissile nuclei and include uranium 235 (U235) and plutonium 239 (Pu239), containing 235 and 239 nucleons respectively.

The first fission reactor was constructed by Fermi in 1942 and this demonstrated the feasibility of nuclear power. He made use of the remarkable property of fissioning nuclei that in addition to energy, principally as energy of motion of the fission fragments, they also emit neutrons, about 2.5 of them on average, and these neutrons can each enter a nearby nucleus and cause it to fission also. Thus provided we do not lose too many neutrons by escape from the reactor or through capture by non-fissile nuclei, we can have a chain reaction, as shown in Fig. 3.1. This occurs provided one or more neutrons from each fissioning nucleus, on the average, causes a further fission. Since the time taken for a neutron to travel from one fissioning nucleus to the next is extremely short (about 10^{-4} seconds), many generations can occur in a small fraction of a second, resulting in a very rapid release of energy. Nuclear reactors are constructed so that less than one of these direct neutrons, on average, causes another fission, so an explosive reaction cannot occur. The chain reaction is sustained by additional neutrons that are emitted some seconds later from some of the fission fragments. The release of energy in a reactor is thus slow and easily controllable. Atomic bombs on the other hand, are designed so that the energy is released as quickly as possible.

Natural uranium is composed of two isotopes, that is nuclei with the same charge Z but with different numbers of neutrons. These are U235 and U238. Only U235 can fission when it captures a slow neutron, and it is present in the very small proportion of about 0.7%. To make a bomb it is necessary to separate the U235 from the U238, a very difficult process since they are both forms of uranium and thus chemically almost identical. A reactor can, however, be made with natural uranium, provided the number of neutrons captured by the U238 is kept at a low level. This is done by using a moderator, whose operation depends on the way the probabilities of capture and fission vary with the energy of the neutron. At the levels of energy they possess when emitted from the fission process, the neutrons are likely to be captured by the U238, and thus removed from the chain reaction. However if their energies can be reduced, they are much more likely to cause fission in the U235. Thus if we slow the neutrons down we can maximise fission and minimise capture. Neutrons may be slowed down by collisions with other nuclei, and lighter nuclei do this most efficiently. We can therefore slow the neutrons down by surrounding the uranium with a light material called a moderator, that does not capture neutrons. Heavy water (deuterium oxide) is the best for this purpose but it is expensive and not enough was available to Fermi in 1942, although it has been used extensively since. Graphite is the next best, since the other light nuclei capture neutrons too readily, and so it was chosen as the moderator for Fermi's reactor. The reactor was essentially a large pile of graphite blocks with rods of uranium in channels passing through the graphite. After they are emitted from a fissioning nucleus, the neutrons enter the graphite and are slowed down so that they can cause further fissions.

During normal working, a reactor is adjusted so that on average exactly one neutron from each fission causes one extra fission, so that the power output remains constant. If this number, called the multiplication factor, has a value below one, the reaction stops, while if it is above one the number of fissions increases rapidly: we say that the reactor has gone critical. The time taken from one fission to the next is so short that if this was all that happens the reactor would be very difficult to control. Fortunately, however, some of the fission fragments also emit neutrons a few seconds later; these are the delayed neutrons. The reactor is then adjusted so that it only becomes critical with these delayed neutrons, and then it is easy to control.

The multiplication factor can be adjusted by using control rods made of material that absorbs neutrons very easily and can be moved in or out of the reactor so that they absorb more or fewer neutrons. To start the reactor, they are moved out slowly until the reactor becomes critical with the delayed neutrons. The power then rises to the desired level, and then the control rods are pushed in again until the multiplication factor is reduced to one, so that the power level remains steady. If we want to increase the power level, the control rods are pulled out slightly so that the multiplying factor rises above one. When the power has reached the desired level the control rods are moved back to their original position so that the multiplication factor is one again. Conversely if we want to reduce the power level the control rods are moved in to make the multiplying factor slightly below one, and then back again when the power has reached the desired level. In an emergency, the control rods can be moved in very rapidly to stop the reaction.

As the reaction proceeds, fission fragments are formed and remain in the uranium rods. As they can absorb neutrons, they slow the reaction down, so that the control rods must be moved out slowly to keep the multiplication factor at unity. Eventually there are too many fission fragments in the reactor, so the uranium rods are removed and replaced by new ones. This can be done continually as the reactor is operating. The fission fragments are highly radio-active and so they must be carefully controlled. This is done by treating the spent rods chemically to separate out the uranium, which can be used again, and the plutonium which is also produced by the neutrons captured by the U238. Plutonium is also fissile and some of it undergoes fission and contributes to the power output. The amount of plutonium produced depends on the type of reactor, but is approximately 200 to 400 kg per year. After this reprocessing, the separated fission fragments constitute nuclear waste, and they are stored as described in Sec. 3.4. This separation of the fission fragments from the uranium and plutonium is carried out in a reprocessing plant such as THORP (Thermal Oxide Reprocessing Plant) at Sellafield in Cumbria. An operating reactor of 1000 MW contains about a thousand megacuries of radioactivity, similar to that released in a 20 kT bomb.

The energy from fission is essentially the energy of motion or kinetic energy of the fission fragments and eventually appears as heat. This heat can be used to generate steam and hence electricity by the same processes which occur in a station burning coal or oil. Many nuclear power stations have now been built and have been operating for years, so it is possible to obtain a realistic estimate of the costs of nuclear power.

There are many types of reactors, and these may be classified by their purpose, their fuel and moderator and the speed of the neutrons when they cause another fission. A reactor may be built for experimental purposes, or to produce fissile material or to generate useful power. The fuel can be natural uranium, or uranium enriched with a higher proportion of fissile uranium isotopes, or with plutonium. The moderator can be any material that slows the neutrons down efficiently without capturing too many of them, and ordinary water, deuterium in the form of heavy water and graphite are often used. Reactors with a moderator are called thermal reactors, whereas those without a moderator are called fast reactors. Fermi's first experimental reactor used natural uranium as fuel and graphite as moderator, and this established the possibility of a chain reaction. Subsequently, large reactors were built to produce the fissile material plutonium 239, which is formed from the non-fissile U238 by neutron capture followed by the successive emission of two electrons. The plutonium differs chemically from uranium, and may therefore be separated from the uranium by relatively easy chemical methods.

It is not possible to construct a reactor with natural uranium and ordinary water because the hydrogen nuclei in ordinary water (light water) combine with neutrons to form deuterons. Too many neutrons are absorbed in this way and the chain reaction cannot develop. If however the uranium is enriched to about 3% fissile U235 or with Pu239 a chain reaction can occur with ordinary water as a moderator. Such reactors are the light water reactors that are widely used to generate nuclear power.

Nuclear reactors require a continual supply of uranium, and so it is important to know how much is available. Uranium ores are very widespread and the known reserves of the richer ones are increasing. As they are used up the less rich can be economically mined. Since the cost of the uranium is only about 5% of the total cost of nuclear power, it can increase several times without appreciable effect on the total cost. This enables poorer ores to be mined and these are very extensive. As a last resort, it is even possible to extract uranium from the sea, which contains about 4500 million tonnes. An experimental plant in Japan produced 10 kg of uranium oxide per year from the sea, though at a cost about ten times the current market price. Further development may reduce the price to \$100 per lb of uranium, but this is still far above the commercial price of \$30 to \$50 per lb. There is thus enough uranium for about a hundred years, but nevertheless there will eventually come a time when the cost becomes so high that thermal reactors will become uneconomic relative to another type of reactor called a fast breeder reactor. (More accurately, these reactors are fast neutron reactors, since the word 'fast' refers to the neutrons and not to the rate of breeding).

Fast neutron reactors can be designed to produce more fissile material than they consume, and hence are referred to as 'breeders'. These reactors contain no moderator to reduce the energy of the neutrons. The neutrons remain fast, and therefore have less chance of causing further fissions. But if the main fuel is Pu239 sufficient fissions occur to sustain the chain reaction and also to convert some of the U238 into plutonium. In one type of fast reactor there is 80% U238 and 20% Pu239 in the interior, surrounded by a blanket of U238. About three neutrons are emitted when Pu239 fissions, and of these one is required to sustain the chain reaction, leaving two to account for losses and to breed more Pu239. Another type of fast reactor uses a blanket of thorium 232, which is converted into the fissile uranium 233. In all these reactors the spent fuel rods are sent to a reprocessing plant to extract the new fissile material. Reprocessing is essential for fast reactors whereas it is optimal for thermal reactors. They thus convert a waste product into a productive fuel.

The power density on a fast reactor is substantially higher than in a normal power reactor so liquid sodium, which cools efficiently without moderating the neutrons, is used to remove the heat. Prototype fast reactors have operated for a number of years in several countries. Their great advantage is that they will enable over 50% of the uranium to be used to generate power, not just the 0.7% of natural uranium. They can either produce or burn plutonium, depending on the neutron density, and can also be used for actinide incineration.

Fast reactors have a strong negative temperature coefficient and so are inherently safe. Due to the low level of corrosion and the ease with which components can be replaced they are likely to have a longer life than thermal reactors, possibly up to seventy years. They also conserve raw material and reduce the quantity of waste products.

The safety of fast reactors was underlined by the accident in 1995 at the Monju reactor in Japan. A cooling circuit leaked two or three tonnes of liquid sodium, which ignites spontaneously on contact with air. This sodium smoulders at the surface, and produces dense fumes of oxide powder, acting as a self-extinguisher. At no time was there any threat to safety.

Since fast reactors burn the uranium 238 remaining in the spent fuel rods from thermal reactors, there is a vast store of energy waiting to be used. It is estimated that the depleted uranium now stored in Britain contains the energy equivalent of all the 500 years' supply of coal now in our reserves. At present it is uneconomic to build fast reactors because uranium is plentiful, but in a few decades this will no longer be so. This period will be shortened if large numbers of thermal reactors are built in the near future. Already the annual consumption of uranium substantially exceeds the annual production. It is thus extremely unwise to abandon the development of fast reactors.

Nevertheless, many countries, including the USA, Britain and France, have severely curtailed or cancelled their fast reactor research programmes. This short-sighted policy is likely to put them at a severe disadvantage compared with Japan when fast reactors eventually become important. Germany's prototype fast reactor at Kalkar, costing £5B was completed but never allowed to operate, a victim of politics. The 1200 MWe French fast reactor Super Phenix at Creys-Malville started to operate in 1985 but has subsequently been closed down by the newly-appointed Environmental Minister, the leader of the Green party in the French Parliament. India has no option but to accelerate its pace towards fast reactors in order to utilise the thorium that is plentiful there. It has had a 13 MW fast reactor operating since 1985, and plans to build a prototype 500 MW fast reactor.

An illustration here might be appropriate: to generate 1000 kWh requires 0.0003 kg of uranium in a fast reactor, compared with 0.025 kg of uranium in a thermal reactor, 350 kg of coal, 250 kg of oil and 280 cubic metres of gas.

3.3 Nuclear Radiations

There are three types of nuclear radiations, called alpha, beta and gamma. Alpha-particles are nuclei of helium, consisting of two neutrons and two protons tightly bound together. They are frequently emitted from naturally occurring radioactive nuclei. Since they are doubly charged they lose energy rapidly and can be stopped by a sheet of paper. Beta rays are electrons of moderate energy; they are more penetrating than alpha-particles and they have only one electric charge and much higher velocities. Gamma rays are electromagnetic radiations and are even more penetrating.

All these nuclear radiations are more or less harmful to living organisms. As they pass through the cells they can break up some of the complicated molecules forming them, releasing highly reactive radicals that can cause more damage. The damage depends on the rate of energy transfer from the radiation to the tissues; the greater the energy loss the greater the damage. Alpha-particles lose energy rapidly, but this means that they are rapidly slowed down and stopped, so they are dangerous only if the source is inside or very near the body. Beta and gamma rays lose energy slowly and so are less dangerous, but they can penetrate the human body and so damage can be caused by sources both inside and outside the body.

Nuclear radiations can easily be detected by very sensitive instruments that will record the passage of a single particle or ray. It is thus possible to detect the presence of extremely small amounts of radioactive substances, and that is why we know so much about how they move through the atmosphere, the oceans and our own bodies. This property has proved extremely useful in many areas of medical research.

In order to make it possible to study radiation damage scientifically, several ways of defining dosage have been devised. This is a complicated subject because, although the amount of energy delivered can be precisely defined, the damage it causes depends on the type of radiation, its intensity and a host of other factors. These ways of defining radiation dosage are described in App. 1.

When considering the effect of nuclear radiations on people, it is also necessary to include the different sensitivities of the different organs of the body. This is done by defining the *rem*, which is the dose given by gamma radiation that transfers 100 ergs of energy to each gram of biological tissue; for other types of radiation it is the amount that does the same amount of biological damage. A new unit, the *sievert*, has now been defined as 100 rem.

Nuclear radiations are often feared because they are unfamiliar and can cause great damage to living organisms without our being aware that anything untoward is happening. The damage only appears afterwards; sometimes very long afterwards, when it is too late to do anything about it. Our senses warn us of many dangers, such as excessive heat and some poisonous gases, and we can take avoiding action. Nuclear radiations are not alone in being invisible; many poisonous gases such as carbon monoxide have no smell, many poisons are tasteless, and we don't know that a wire is live until we touch it and receive an electric shock.

Nuclear radiations are feared not only because they are invisible, but because they are unfamiliar. Ironically enough, when they were first discovered in the context of X-rays, luminous radium salts and radiotherapy, they were hailed with enthusiasm as great benefits to mankind, and to some extent rightly so. As a special recommendation, bottles of health-giving mineral waters were advertised as radioactive. It was only much later, when more was known about the damage that nuclear radiations can cause to living cells, and especially when pictures were released of the radiation damage to the victims of Hiroshima and Nagasaki, that the public image of nuclear radiations switched to one of fear.

Undoubtedly this reaction has gone too far. Nuclear radiations are indeed dangerous in large amounts, but so are fire and electricity. Properly used, nuclear radiations have numerous beneficial applications in medicine, agriculture and industry. Like so many of God's gifts, they can be used for good or evil.

When assessing the dangers of nuclear radiations it is important to realise that they are not new. They did not first enter the world with the experiments of Henri Becquerel and Madame Curie. They have been around on the earth since the very beginning. Many rocks and minerals, such as the pitchblende that Madame Curie refined to produce the first samples of radium, are naturally radioactive, and some emit nuclear radiations all the time. The earth is bathed in the cosmic radiation that comes from the depths of outer space. These rays are passing through our bodies all the time. The human species has evolved through millions of years immersed in nuclear radiations. They may even have helped evolution by causing the occasional beneficial mutation.

Reviews of radiation hazards by the National Academy of Science in the USA and by the United Nations give an estimate for cancer of about one hundred cases per million person-rem exposure. This means, for instance, that if a million people are exposed to an extra rem during their lifetimes then a hundred extra cases of cancer can be expected over and above the 200,000 cases normally expected for this number of people. The effects of smaller or larger doses are estimated in a similar way.

It is often rather a delicate matter to decide whether the benefits of medical irradiation outweigh the hazards. For example, X-rays can detect cancer early enough for effective treatment, but they can also cause cancers. A programme of mass X-rays will therefore cause some and help to remove others. The only way to decide this question is by statistical studies of the number of tumours caused and the number cured. A particularly detailed study of stomach tumours showed that for young people the dangers outweigh the benefits, whereas for older people the opposite is the case.

This illustrates the essential truth that the only way to assess the hazards of nuclear radiation is to express them numerically. Although we cannot detect nuclear radiations with our unaided senses, there are sensitive instruments that will do this for us, by measuring the intensities of radiations of various types that we receive naturally and through medical treatment. The figures vary in many ways. The natural radiation depends on the proximity of radioactive minerals; these are widespread in small amounts but are more often found in igneous than in sedimentary rocks. This cause alone can give values of the background radiation varying from one place to another by factors of two or three, and much higher factors in a few locations. The cosmic radiation is attenuated progressively as it comes down through the atmosphere, so the higher up we go, the more cosmic radiation we receive. If we go deep into the earth, the cosmic radiation is reduced almost to nothing, but the radiation due to rocks and minerals may increase greatly, depending on the geological formation. This again can account for large differences between one place and another. Finally quite a large dose of radiation can be received medically. Some figures for the average dose of radiation from various natural sources are given in Table 3.1 and illustrated in Fig. 3.2.

These doses are all well-known, familiar and accepted without a qualm. When we consider the radiation from the nuclear power industry it must like-wise be expressed in numerical form to make possible a proper comparison.

Table 3.1. Radiation Exposure in the United Kingdom.

Source		Exposure in Millirem per year	
Cosmic Radiation	⎫	31	⎫
Terrestrial gamma rays	⎬ Natural	38	⎬ 186
Internal Radiation		37	
Radon Decay Products	⎭	80	⎭
Medical Irradiation	⎫	50	⎫
Fallout from bomb tests		1	
Miscellaneous sources	⎬ Man-made	0.8	⎬ 53
Occupational Exposure		0.9	
Disposal of Radioactive Wastes	⎭	0.3	⎭

(*National Radiological Protection Board Bulletin*, March 1981; *Atom*, May 1981).

Radiation exposure of the UK population

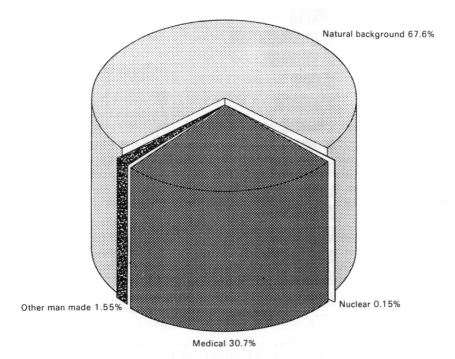

Natural background 67.6%

Other man made 1.55%

Nuclear 0.15%

Medical 30.7%

Fig. 3.2. Percentage contributions to the total radiation exposure of the population in Britain. (*Atom*, January 1980 p. 4)

The main radioactive hazard from a nuclear reactor comes from the fission fragments. These are intensely radioactive and continue to emit nuclear particles and radiation for a long time.

In normal operation, a nuclear power station releases only very small amounts of radioactivity into the atmosphere. These amounts are even less than the radioactive emissions from coal power stations.

To put them into perspective they must be compared with other sources of radiation. We are constantly exposed to the same types of nuclear radiations from radioactivity in the earth and from the cosmic radiation that enters the earth's atmosphere from outer space. This natural radiation amounts to about 1.86 millisievert (mS) per year, of which 0.31 comes from the cosmic radiation,

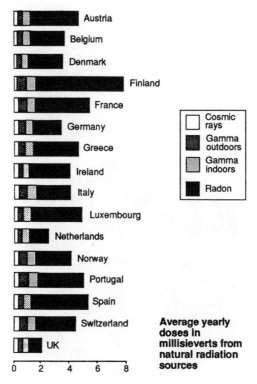

Fig. 3.3. The average yearly doses from the natural radiation sources in European countries (National Radiological Protection Board (NRPB), 1993).

0.38 from the soil and the air, 0.3 from the radioactive material in our own bodies and 0.8 from the decay products or radon emitted from radioactive rocks. In addition we receive on the average about 0.5 mS from X-rays and other forms of medical treatment and about 0.01 mS from the fall-out from atomic bomb tests. The radiation from nuclear power reactors is kept carefully under control and is currently about 0.3 microsieverts per year. The relative proportions of these contributions to the radiation we receive are shown in Fig. 3.2.

All these radiation levels are far below those that could cause any sort of illness. We know this because in some parts of India and Brazil the natural background is over ten times the average quoted above, due to the presence of radioactive rocks, and even in these areas the population shows no signs of additional radiation damage that could be attributed to this extra natural radiation. Professor Fremlin from Birmingham has estimated that the radiation dose received in Britain due to the nuclear power stations reduces our average life expectancy by one or two seconds. The radiation dose is similar to that from coal power stations, since all coal is contaminated by uranium.

There is widespread public anxiety about the effects of nuclear radiations, particularly concerning the genetic effects and the cases of childhood leukaemia near nuclear plants. The genetic effects in first generation offspring are abortions, still-births, congenital defects, infant mortality, reduction in birth weight and changes in sex ratios. The children of the survivors of Hiroshima and Nagasaki, who all received massive doses of radiation, have been studied in detail. These children showed no differences from normal children, so fortunately there is no evidence for genetic damage. Seven cases of childhood leukaemia occurred in Seascale in Cumbria near the nuclear reprocessing plant at Sellafield from 1955 to 1983. This number seemed to be much greater than would be expected by chance, and received much publicity. It was, however, very difficult to understand how these cases could be blamed on Sellafield, since the amount of nuclear radiation released is far smaller than the natural background.

To estimate the biological damage that can be caused by a particular dose of radiation we must know the relationship between the two quantities. The difficulty is that the doses that cause measurable damage are hundreds of thousands of times larger that the doses received by people living around nuclear installations. We know about the radiation damage due to the massive doses of several hundred rem received by the victims of Hiroshima and Nagasaki, and in the treatment of certain conditions like rheumatoid spondylitis. In assessing the radiation hazards of nuclear power, however, we are concerned with doses measured in thousandths of a rem, which produce no observable effects. Can we assume a simple linear relationship between dose and damage, so that the probability of contracting cancer is proportional to the dose? There is little direct evidence for this, and some contrary evidence, but it seems to be the safest assumption to make, and is the one adopted for many radiation standards. It is the basis of calculations made by environmental organisations who want to emphasise the hazards of nuclear power. It is however the experience of medical studies on the effects of a wide range of substances that are certainly hazardous in large doses that there is no evidence for a proportionate hazard at extremely small doses. This is hardly surprising, since the body has an innate capacity to repair damage; it is only when the defences of the body are overwhelmed by a massive dose that harm occurs. Thus a dose received over a long period is less harmful than if it were received all at once. There is indeed evidence that there is a threshold dose below which there is no damage at all. This has been summarised by the French physiologist Claude Bernard who remarked that at high enough levels everything is toxic, whereas at low enough levels nothing is toxic.

There is even some evidence that small doses of radiation are beneficial. Thus a very extensive study made by Frigerio *et al.* at the Argonne National Laboratory in 1973 compared the cancer statistics for the white population of the USA from 1950 to 1967 with the average natural background for each State. They found that the seven States with the highest natural background had the lowest cancer rates. Unless there is some other reason for this unexpected result it implies that the chance of contracting cancer is *reduced* by about 0.2% per rem. Similar effects have been found in studies of the irradiation of animals. Thus the irradiation of mice by gamma rays increases their survival rate by one week per rem. The irradiation of salmon eggs increased the number of viable eggs, and the rate of return of the adult fish to their birthplace to breed. Experiments by Dr. Latarjet of the Institut Curie in Paris showed that the number of mutations produced by a large radiation dose is reduced if the specimen is exposed to a much smaller dose six hours previously. This suggests that the smaller dose stimulated the cell repair mechanism.

The belief that radiation damage is linearly proportional to the dose, with no threshold, has led to exposure regulations which resulted in exaggerated fears of radiation and costly consequences. Thus Morris Rosen, the coordinator for environmental affairs at the International Atomic Energy Agency, has stressed that radiation exposure regulations should bear a sensible relation to the everyday exposure to the background radiation and its variations from place to place. At Chernobyl four hundred thousand people were relocated on the basis of non-threshold regulations although the total dose they received from Chernobyl and background is less than the background radiation alone in many other places such as Finland and France, as well as Devon and Cornwall in the United Kingdom. Very costly changes have had to be made in mining methods to meet the latest regulations of the International Commission for Radiation Protection. Using the risk factor of 4×10^{-2} per Sv for fatal cancers, the cost to save the fifteen or twenty years of life normally lost in a case of lung cancer was in the range \$M20 to \$M150.

In favour of the concept of a threshold dose, it can be argued that the passage of a single nuclear particle through a cell, the lowest possible dose, can cause DNA double strand lesions. Such lesions however occur naturally at the rate of about ten thousand per cell per day, whereas exposure to radiation at the current population exposure limit would cause only two lesions per cell per day. The radiation-induced lesions are thus insignificant compared with those occurring naturally.

Nevertheless there remain the cases of leukaemia around Sellafield, five in all, well above the number expected statistically. There are however other cases of leukaemia clusters around nuclear plants that were never built, or built only after the cases had occurred. It remains important to see if there is an explanation for the Sellafield cases.

A possible explanation is that they are connected with the influx of many people into a relatively isolated rural community, as occurred in Seascale near Sellafield. The cases then could be due to some virus infection, as proposed by Dr. Kinlen. This hypothesis can be tested by seeing if the effect occurs in the case of similar population movements involved in the construction in rural areas of factories that have no connection with the nuclear industry. This possibility has been examined for several such cases in the north of Scotland, and for children evacuated from London in 1941 during World War 2, and indeed an excess of leukaemia cases was found in each case. It still remains to establish the mechanism for this effect.

Another possible mechanism was suggested in 1987 by Gardner, who postulated that the children developed leukaemia as a result of their father's exposure to nuclear radiation. He collected statistics that showed a significant correlation between paternal radiation dose and leukaemic children. This led to several Court cases in which families sought compensation from British Nuclear Fuels, the company operating the plant.

The Gardner hypothesis has such serious implications for the nuclear industry that many further investigations were made. These were on the actual process whereby paternal irradiation could lead to childhood leukaemia, the observations of leukaemia in the children of survivors of the atomic bombing of Japan, and more extensive studies of leukaemia around nuclear plants. The results of these studies have now been summarised by Sir Richard Doll, Dr. H. J. Evans and Dr. S. C. Darby in *Nature* (367.678.1994). They conclude that the Gardner hypothesis is wrong.

The possibility that nuclear irradiation could cause a gonadal mutation leading to childhood leukaemia can be studied using data on genetically-determined leukaemia. The detailed statistical knowledge shows that there may be a recessive mutation that could contribute to a number of observed cases. However, "it effectively excludes any major contribution from the type of mutation that would be required to account for the appearance of the Sellafield cases in the first generation, namely a dominant mutation with a high degree of penetrance".

Studies by Neel and colleagues of "the children of atomic bomb survivors, including more than 1500 born to parents who received a gonadal dose of one sievert or larger, revealed no clearly increased frequency of mutations". These doses are far higher that those received by the Seascale workers.

Further studies were made of all the leukaemia cases in people under 25 years of age in 1958–90 born after 1958 in Scotland and a part of north Cumbria near the Scottish border, and of all children under 15 born near five nuclear installations in Ontario. They found that "neither set of results supported the probability of a hazard from the father's occupation". Several other studies have reached the same conclusion.

Thus the authors conclude that "the association between paternal irradiation and leukaemia is largely or wholly a chance finding". They note that there appears to be "small but real clusters of leukaemia in young people near Sellafield, and some other explanation for them needs to be sought".

This highly authoritative study should finally lay to rest the fears of nuclear radiations from plants like Sellafield, but whether it will or not depends largely on the mass media. The presence of leukaemia clusters, and particularly the Gardner hypothesis, has been widely publicised by organisations opposed to nuclear power. This has encouraged families with children suffering from leukaemia to seek compensation, but when the scientific evidence was laid before the court, the judgement inevitably went against them.

There is also some concern about the possibility of radiation doses received by people who eat sea food from the Sellafield region. Studies have shown that the few people who eat very large amounts may receive an extra annual dose of 0.35 mS. Those living near Sellafield may receive an extra dose of 0.25 mS. This is to be compared with the average annual background dose of 2.2 mS per year and about 8 mS in Cornwall.

Similar studies in other countries gave the same results. Thus a French study by the Institut Gustave Roussy in 1990 recorded the leukaemia rate around the reprocessing plant at La Hague, including fuel cycle and reactor operation and four nuclear power stations. Around these sites in the period 1968–1987 for the age group 0 to 24, there were 55 leukaemia deaths compared

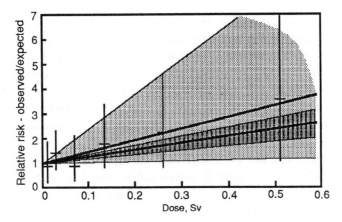

Fig. 3.4. The relative risk of leukaemia, expressed as the ratio of observed to expected cases, as a function of radiation dose. The heavily-shaded band gives the results found by the International Commission on Radiation Protection and the wider lightly-shaded band those of the National Radiological Protection Board (NRPB) (*Nuclear Issues*, 1992).

with 62 in similar control areas and 67 from the national mortality statistics. A US study of 60 major nuclear installations found that the leukaemia mortality was lower in the nuclear areas from the time that the plants came into operation in 1984.

Detailed studies by the National Radiological Protection Board and the International Commission on Radiation Protection of the life histories of thousands of workers have shown "that there is no evidence that radiation workers have cancer mortalities which are higher than those for the general population". If anything, the statistics showed smaller rates. The relative risk expressed as the ratio of observed to expected cases is plotted as a function of dose in Fig. 3.4. It is notable that the result is consistent with zero risk and with the presence of a radiation threshold below which no radiation damage occurs. A study of workers in the UKAEA in 1985 showed that the death rates from cancer were 22% lower than the national average, and 14,347 workers at Sellafield had an average mortality rate from all causes that was 2% less than the national average. A study of 21,358 men who took part in the UK atomic bomb tests in Australia and the Pacific showed no detectable effect on their life expectancy or on the incidence of cancer or other fatal diseases.

In the UK, the Committee on Medical Aspects of Radiation in the Environment studied the radiation levels around the Atomic Weapons Research

Establishment at Aldermaston, Burghfield and AERE, Harwell and found no evidence of radiation of sufficient intensity. This is not surprising, since NRPB estimated the additional doses to be respectively 0.0017%, 0.0000003% and 0.032% of the natural background radiation. A similar comparison of the radiation levels near coal and nuclear power stations by the NRPB found 0.00034 mSv at 5 km. from a coal power station, and 0.000063 mSv at the same distance from Hinkley Point C nuclear power station. The annual background radiation is about 2.5 mSv.

Professor Fremlin has estimated that during the next thousand years a constant population of a million people will receive from natural sources a total collective dose of 1500 million man-rem. This is far greater than the tiny addition due to man-made sources.

The concern about nuclear radiation has diverted attention from other threats to our health. Radiation is responsible for only about 1% of diseases worldwide, and most of this comes from the natural background and from medical uses. The nuclear industry is responsible for less than 0.01%. The vast sums spent to reduce this still further could be spent far more effectively on simple disease prevention.

It is greatly in the public interest that these matters should be treated as objectively as possible, taking full account of the scientific evidence. This would avoid much unnecessary anxiety, and enable the best decisions to be taken concerning our future energy supplies.

Nuclear power stations produce radioactivity in the form of fission fragments. This constitutes nuclear waste, and the methods used to get rid of this waste safely are described in the next section.

3.4 Nuclear Waste

The main hazard due to the operation of nuclear reactors is from the waste they produce, in particular the fission products that remain in the fuel rods. As the uranium is burned in the nuclear reaction, the fission products accumulate until they prevent the reactor from working. To avoid this, spent fuel rods are continually removed from the reactor and replaced by new ones. The fuel rods go to a reprocessing plant, where the uranium and plutonium are separated and used to make new fuel rods. The fission products are highly radioactive and emit various types of particles and rays. There are alpha- particles that are stopped by a sheet of paper, beta particles that are stopped by a thin metal sheet, and gamma rays that need an inch or two of lead to stop them. These nuclear radiations, as they are called, can damage human tissue, and a very high dose can cause loss of appetite, illness and even death.

It is therefore essential to ensure that the fission products do not escape into the atmosphere or enter the food chain. This is done by first storing them in tanks above ground for fifty to a hundred years to allow most of their radioactivity to decay away. Fission products are a mixture of many different types of nuclei. Most of them are unstable, and their decay half-lives (the half-life is the time that a sample takes to decay to half its initial intensity) vary over a very large range from thousands of years to a small fraction of a second. The nuclei with short half-lives decay quickly and the radiations they emit produce a lot of heat. This heat must be continually removed to prevent the liquid boiling. When the liquid containing the fission products has cooled down, it is reduced to solid form and fused to form an insoluble glassy or ceramic substance. This is encased in stainless steel cylinders, surrounded by concrete and buried deep underground in a stable rock formation. There is then no chance that the fission products will escape and cause harm. A detailed study sponsored by the European Community and completed in 1988 showed that even after millions of years the radioactivity that could be released into the environment is far less than 1% of the natural background level. Eventually,

Table 3.2. Estimated costs of waste disposal in U.S. mills per kWh of electricity production.*

NUCLEAR		Once through fuel cycle	Reprocessing fuel cycle
Uranium mill tailings		0.03 to 0.06	0.03 to 0.06
Low level waste		0.09	0.09
Spent fuel, high-level &	Pre-disposal	0.45	0.73
transuranic waste	Disposal	0.23 to 0.38	0.20 to 0.40
	TOTALS	0.80 to 0.98	1.05 to 1.28
COAL			
Mining and cleaning (solid)		0.004 to 0.27	
Sulphur oxides		2.21	
Particulates		0.66	
Bottom ash and fly ash		0.45	
Scrubber sludge		0.41 to 0.69	
	TOTAL	3.73 to 4.28	

*Report of U.S. General Accounting Office (*Nuclear Issues*, August 1982).

over the years, the radioactivity of the fission products will decay until it is similar to that of the surrounding rocks. This is a very well-understood process for dealing safely with nuclear waste. It has been estimated that the cost of dealing with nuclear waste adds 0.02 p/kW to the cost of electricity. The cost of waste disposal from coal power stations is three times higher, as shown in Table 3.2.

The amount of radioactive waste from nuclear reactors is not large. Every year, a reactor produces about four cubic metres (cm) of high-level waste, 100 cm of intermediate level waste and 530 cm of low-level waste. The total amount of high-level waste produced by the United Kingdom nuclear programme from 1956 to 1986 is less than about 2000 cm, about the same volume as an average house. This is a minute quantity compared with the huge amounts of hazardous chemical wastes produced by the manufacturing industries. The high-level nuclear waste can be further treated chemically to separate the actinide nuclei that contain the dangerous long-lived alpha-particle emitters. These have an even smaller volume and can be put in a very secure depository or destroyed as described below.

To illustrate the quantities involved in waste disposal, a nuclear fuel pellet having about half the volume of a cigarette will generate enough energy to supply the electricity needs of a family of four for a year. To obtain the same amount of electricity from other sources would require a ton of oil, two tons of coal, five tons of wood or 30,000 cubic feet of gas. After reprocessing, the waste from the original pellet would be about the size of a pinhead.

A new way of treating nuclear waste is now under discussion. This is to bombard the waste with high-energy proton beams that break up the radioactive nuclei into less active ones. Alternatively, the proton beam is allowed to hit a target of some heavy element like uranium; this produces a shower of neutrons that can also be used to disintegrate the waste nuclei. Another possibility is to design a sub-critical reactor using thorium as fuel that can become critical with extra neutrons generated in the way just mentioned. Such a reactor can be used to burn nuclear waste and to generate electricity in the usual way. Thorium is particularly convenient as it is far more abundant than uranium. The advantage of these methods of treating waste is that the radioactivity is actually destroyed, instead of being put in a safe place. It is however likely to be rather more expensive.

There are other types of radioactive waste that come from a wide range of industrial and medical applications of radioactive isotopes. These are used in wrist watches and smoke alarms, and as tracers in medical diagnosis and medical research. The radioactivity of these wastes is low compared with that of the fission fragments, and they can be safely disposed of by burying them in a surface trench or by encasing them in concrete and sinking them deep in the ocean.

Scientific studies of ocean disposal have shown that the risks are vanishingly small. Nevertheless, the London Dumping Convention voted to impose a moratorium, providing yet another example of a decision imposed by political rather than by scientific considerations. Instead, the waste is now buried in the earth, where it is almost as safe. The opponents of nuclear power now reject not only waste disposal but also the scientific studies undertaken to find the best ways of doing it. It may be remarked that the seas already contain 4,000 million tons of uranium and other radioactive elements, vastly greater than the amounts we want put there.

The two major sources of energy for the future, coal and nuclear, thus produce very different types of waste. That from coal power stations goes into the atmosphere as well as being produced in solid form. Nuclear power stations, however, discharge almost no waste into the atmosphere; all the waste is in the radioactive fission fragments. The solid waste from both types of power stations can be safely dealt with, but the atmospheric waste due to coal and other fossil fuel power stations is a most intractable problem with most serious consequences for health and for the environment.

The net result of the nuclear power industry is to make the earth less radioactive than before. This is because the half-life of the uranium that is burned is much greater than those of the waste products of fission. Taking into account the relative biological effectiveness, the fission products emit less radioactivity than natural uranium after about five hundred years.

3.5 Applications of Nuclear Power

In addition to the generation of electricity and the supply of direct heat to surrounding houses, there are several other applications of nuclear power. One of the more important is desalination. In many parts of the world there is not enough water, and this can have catastrophic consequences. We are familiar with the pictures of starving people in Africa where drought has caused the crops to fail. Even in Britain, surrounded by water, there are periods of more or less serious water shortage. In places near the sea, the obvious solution is to use sea water, but first the salt must be removed. This may easily be done by boiling the water and condensing the steam, but this requires large amounts of heat, and so is relatively costly. The heat can be obtained from oil or coal, but this increases the production of carbon dioxide and hence, to the extent that the greenhouse effect is operative, it increases the drought. The obvious solution is to use nuclear power plants to desalinate the water, and this possibility is being actively studied. It would be useful to convert ageing nuclear plants to sea water distillation because they could then be run at a lower temperature than that required for the generation of electrical power.

Another possible application is to use nuclear explosions for large-scale rock blasting, such as that required during the construction of the Panama canal. While this would certainly be effective, the radioactive contamination makes this application totally impracticable.

A very practical and useful application of nuclear radiations is the irradiation of food to preserve it from insect pests. This has been extensively studied and approved in 1983 by the United Nations Food and Agriculture Organisation and by the World Health Organisation. It is now permitted in 36 countries for about forty foods including spices, grains, chicken, meat, fruit and vegetables. There has been strong opposition to food irradiation from some consumer groups, influenced by fears of radiation. It has been said that irradiation masks the bad smell and taste of spoiled food, but this is not the case. It is the alternative methods of disinfecting spices that are dangerous, such as fumigation by ethylene dibromide, which is carcinogenic.

The need for such irradiation is shown by the increasing frequency of cases of food poisoning, amounting to many tens of thousands of cases worldwide every year. For example, in February 1993 the French Ministry of Health reported the deaths of 63 people from eating listeria-contaminated pork. In the same year, three deaths and over four hundred cases of food poisoning were caused in the USA by undercooked hamburgers. The objections to food irradiation are just the same as those made a century ago to oppose the pasteurization of milk. This opposition was eventually overcome in about fifty years, but meanwhile very many lives had been lost.

Nuclear power has proved very useful in satellites. The Voyager 2 spacecraft is equipped with a radioactive thermonuclear generator that uses the heat generated by the alpha decay of plutonium. Initially it had an output of 150 W and after twelve years it is still producing a large fraction of that power. This enables the 22 W transmitter to send data back to the earth that should still be detectable until 2020, when the spacecraft will be far outside the solar system. No other energy source could provide power for so long.

3.6 Fusion Power

We already obtain all our energy directly or indirectly from a fusion reaction, for that is how the sun generates its heat. The fusion reaction also takes place in the hydrogen bomb. To make it into a useful source of energy a way must be found to release the energy in a controllable way.

The energy of the type of fusion reaction that is practicable on earth is obtained when isotopes of hydrogen combine together to form more stable particles. The most promising reactions are:

DD: Deuteron + Deuteron → Helion + Neutron + 3.3 MeV.
     and   → Triton + Proton + 4.0 MeV.
DT: Deuteron + Triton   → Alpha-Particle + Neutron + 17 MeV.

The DT reaction is preferred because it has the higher energy yield, and also because of its lower ignition temperature (about a hundred million degrees) compared with that of the DD reaction (about three hundred million degrees).

Current research is therefore devoted to designing a fusion generator using this reaction. It does however require radioactive tritium, which has to be made specially using lithium, whereas deuterium is found in ordinary water. Thus in the long run, deuterium has considerable advantages.

The main difficulty in getting these reactions to take place is that deuterons and tritons are electrically charged and so repel each other. The attractive nuclear forces only come into play when the nuclei are very close to each other, and this requires that they collide at very high energies. Since it is impracticable to aim each one individually, the only way to achieve this is to heat the hydrogen isotopes to a very high temperature of hundreds of millions of degrees. This is done in the hydrogen bomb by using the fission reaction as a detonator.

In the hydrogen bomb the fusion energy is released explosively. If we want to harness this energy in a controllable way we must find a way of releasing the fusion energy slowly. At the very high energies necessary for the fusion reaction the hydrogen is in the form of a plasma, with all the electrons stripped away from the atoms. For a fusion reaction to occur the plasma must be kept together at a high temperature long enough for the fusion to take place. Since the hydrogen nuclei are charged, the plasma can in principle be contained by applying a magnetic field. In such a field a charged particle follows a curved trajectory and one might hope to design a magnetic field so that the particles go round closed orbits and never escape. We would then be well on the way to controlling the fusion reaction.

Many attempts have been made to design 'magnetic bottles' to contain the plasma, but it was found always that the instabilities developed, allowed some of the particles to escape. A promising way to overcome this difficulty is to use a container in the shape of a torus or 'doughnut', and to put it in a magnetic field generated by currents passing through coils wound around the torus. Such fusion devices have been built at Princeton in the United States, at Culham in Britain as well as in Russia and Japan.

To produce a fusion reaction in such a device, a high vacuum is made in the torus and a mixture of deuterium and tritium injected into it. This mixture is then heated electrically so that the hydrogen nuclei collide violently with each other. They are prevented by the magnetic field from colliding with the walls of the torus. The energy is produced when a deuteron and a triton combine to form an alpha-particle, releasing a free neutron. The 17 MeV of energy released is mostly carried away by the neutron. The neutrons are unaffected by the magnetic field and easily penetrate the walls of the torus and enter the surrounding blanket of absorbing material. The neutrons give up their energy to the absorber, which is raised to a high temperature. This can be used to heat steam and hence to drive an electricity generating plant.

The main difficulty is that so far it has not been possible to confine the plasma long enough for the reaction to give a net energy gain. Considerable progress has been made in confining the plasma for longer and longer times, and it is hoped that improved designs will be successful.

The fuel required for fusion reactors is a mixture of deuterium and tritium. Deuterium is present in ordinary water at a concentration of 1 in 6,000, so there is a practically unlimited supply. Tritium is unstable, with a half life of about twelve years, so it has to be manufactured. This can be done by bombarding lithium with neutrons, and this could take place in an absorbing blanket around the reactor. In this way is could produce all the tritium it needs, just like a breeder reactor. The lower ignition temperature means that it is likely that the first successful fusion reactor will use a deuterium-triton mixture, in spite of the need to make the tritium. Later on, when new designs can reach higher temperatures, the DD reaction can be used. We will then have enough energy to last for billions of years, because the fusion energy in a pound of water is ten times that in a gallon of petrol.

There are serious hazards associated with the operation of a fusion reactor, particularly because of the very high energies stored in the magnetic field, around 10^{11} Joules. There is also the danger of a lithium fire and of the escape of radioactive tritium.

It is also possible in principle to initiate a thermonuclear reaction with electron or ion beams focused from several directions on target DT pellets. Another type of fusion reactor depends on the ability of lasers to deliver enough energy to small deuterium-lithium pellets to initiate the fusion reaction. In one of the proposed designs, a system of mirrors splits the beam from a powerful laser and focuses them from different directions on to a deuterium-lithium pellet. The initial effect of the beams on the pellet is to vaporise some of the surface and to set up an imploding shock wave that helps the fusion reaction. Some experiments have shown that fusion can be achieved in this way, but so far the energy required to power the lasers vastly exceed that produced by fusion.

When a design has been found that produces a net gain in energy, it will be time to design a fusion reactor for full-scale energy generation. Experience with fission reactors shows that it takes about twenty years from the establishment of the technical viability of a reactor system to its large-scale implementation for power production. Fusion power is so important that research is being pressed forward rapidly, but until fusion power stations are in operation it should not be included in any discussions of our future energy sources.

References

Boyes, E. (Ed), *Shaping Tomorrow*, Home Mission Division of the Methodist Church, 1981.

Cohen, B. L., *Nuclear Science and Society*, Anchor Books, New York, 1974.

Cottrell, A., *How Safe is Nuclear Energy?* Heinemann, 1981.

Gowing, M., *Britain and Atomic Energy 1939–1945*, Macmillan, 1964.

Greenhalgh, G., *The Necessity for Nuclear Power*, Graham and Trotman Ltd., London, 1980.

Grenon, M., *The Nuclear Apple and the Solar Orange*, Pergamon Press, 1981.

Hafele, W., *Energy from Nuclear Power*, *Scientific American* September 1996, p. 91.

Hoyle, F., *Energy or Extinction? The Case for Nuclear Energy*, Heinemann, 1977.

Hoyle, F. and Hoyle, G., *Commonsense in Nuclear Energy*, Heinemann, 1980.

Kinlen, L. J. and John, S. M., *Cancer Research Epidemiology Unit*, Oxford; *The Lancet* 19 December, 1980.

Marshall, W., *The Disposal of High-level Nuclear Wastes*, Atom October 1981.

Pentreath, R. J., *Nuclear Power, Man and the Environment*, Taylor and Francis, London, 1980.

Wade, B., *Radiation and Nuclear Power*, Atom November 1981.

4

THE SAFETY OF NUCLEAR REACTORS

4.1 Introduction

Nuclear reactors are designed to be safe, which means that it is very unlikely that anything will go wrong, but also that if something does go wrong there are no serious consequences or danger to the public. Especially important are what are called passive safety features that ensure that the reactor responds automatically, without any intervention by the operators, to correct any malfunction. One of these passive safety features is what is called a negative temperature coefficient; this ensures that if the temperature of the reactor rises for any reason, the reactivity of the reactor automatically falls and reduces the temperature to the correct level. A serious design fault of the Chernobyl reactor was its positive temperature coefficient, that had the opposite effect. Then there are the control rods, held above the reactor by electromagnets so that if the electrical power fails they are immediately released and fall into the reactor and stop the reaction. There are massive flywheels on the pumps that circulate the cooling water so that they continue to operate for some time if the power fails. Emergency cooling is provided by large vessels of boronated water containing nitrogen which automatically discharge into the reactor if the pressure drops.

Additional safety features proposed for new reactors include large tanks of water above the reactor that can provide replacement cooling water, and automatic systems that can detect and remove any hydrogen that may be released. To prevent the escape of radioactivity the whole reactor is inside a 14-inch thick steel pressure vessel and it is surrounded by a massive concrete containment building.

Detailed studies have been made of all the conceivable malfunctions of nuclear reactors, so that they can be anticipated and allowed for in the design. Thus it is conceivable that a cooling water pipe may fracture, release the coolant and thus allow the temperature of the reactor to increase.

If a nuclear reactor is badly designed, or if it is operated in a careless way, accidents can occur, the power level may rise and radioactivity may be released directly into the atmosphere. A nuclear reactor cannot explode like an atomic bomb because even if the core went critical it would blow itself apart before a true atomic explosion could take place.

Most reactors are designed so that they are thermally stable, and if for any reason they overheat, the reaction is stopped by the control rods. Even after this is done, the heat continually given out by the fission fragments in the fuel rods could melt the uranium and then the steel containing vessel. If this fractures, large amounts of radioactivity would be released. If the reactor building is also damaged, the radioactivity could be released into the atmosphere, resulting in a serious health hazard.

Normally, such a rise in the temperature of the reactor core is prevented by the circulation of cooling liquid, but this may be interrupted if the pipes are broken. To cope with such an emergency there are usually duplicate cooling systems, and provision to inject more cooling fluid whenever the pressure in the primary system falls below a safe level.

As in any industrial enterprise there have been many minor accidents to nuclear reactors, usually involving leaks of radioactive liquids. In most cases these were easily dealt with, and no radioactivity escaped from the reactor building. Several accidents were more serious, such as the sudden release of energy in the early Windscale reactor in 1957 due to the Wigner effect. This led to some release of radioactivity into the atmosphere.

A serious accident occurred at the military plant at Kyshtyn in the Southern Urals in 1957. The cooling system of a storage tank containing high-level radioactive waste failed, and this led to a chemical explosion. As a result, radioactive material containing 7.4×10^{16} Bq was spread over an area about 800 km long and 8 km wide. There were no fatal casualties and no observable medical effects.

Other accidents were much more serious, and showed how poor design and operator error can lead to disaster. Those at Brown's Ferry and Three Mile Island severely damaged the reactor but caused little harm outside the reactor building. The resultant publicity, especially about the Three Mile Island accident, caused great harm to the reputation of nuclear power.

By far the worst accident was the one at Chernobyl which exposed many people to intense radiation and contaminated a large area with radioactive material. This was a major disaster that has caused even more damage to the nuclear power programme.

These three accidents are described in the following sections.

4.2 Brown's Ferry

All the accidents illustrate the results of faulty design and faulty operating procedures. At Brown's Ferry the trouble started when an electrician was testing for air leaks in the cable room beneath the main control room. He used the standard test for leaks, a lighted candle, whose flame is sensitive to air pressure. There was indeed a hole, and because the air in the reactor containment building is kept at a lower pressure than outside (to prevent the escape of radioactivity) the flame was sucked into the hole. There it set fire to strips of polyurethane foam inside the hole. This in turn set fire to the insulation of the electric cables, many of them controlling the reactors safety system. This made it very difficult for the reactor engineers to cool the reactor down. Eventually they succeeded in doing this sixteen hours after the fire had started by using the auxiliary water supply. The fire itself was out of control for over seven hours and damaged sixteen hundred control cables causing millions of dollars damage. After that experience flammable material is never used for cable insulation and there are two independent sets of cables to supply power to the safety equipment. Fortunately there was no release of activity in this accident.

4.3 Three Mile Island

The accident at Three Mile Island was of a different character. The first thing to happen was a breakdown of the pumps circulating water in the secondary cooling system. The water in this system circulates near the water from the primary system that comes from the reactor and removes some of its heat. The standby cooling systems should then have come into action, but the valves were closed, which was not known at the time. Since it was no longer being cooled, the water in the primary system heated up. This automatically shut down the reactor, but the radioactive core still continued to emit heat. The pressure in the primary water system was then reduced by a valve above the reactor, which failed to close. Additional water was then pumped in automatically to replaced the primary water. The operators then misinterpreted a dial reading and, thinking that it indicated that there was no vapour phase above the cooling water in the pressuriser, stopped the emergency water supply for fear of filling the system with liquid which on expansion would have burst the containing vessel.

The top of the reactor core then became uncovered and its temperature rose rapidly. The fuel rods buckled and cracked, and the zirconium casing reacted with the water to give free hydrogen, which collected in a bubble at the top of the reactor vessel. There was then some concern that the hydrogen would react explosively with any oxygen that might be present from the water molecules, bursting the reactor vessel and perhaps also fracturing the containment building, allowing a massive release of radioactivity. Actually, there was never any danger of this happening because the zirconium combines with any oxygen that might be present, but this was not realised at the time and several alarming announcements were made to the public before the reactor was eventually brought under control.

Only a small amount of radioactivity was released, enough to give people living within fifty miles the dose (about one millirem) that they would receive in one day from natural sources. A US court has recently dismissed two thousand damage claims for adverse medical effects brought against the company owning the plant. The major health effect was the distress caused by the spread of alarming and inaccurate information.

There were some widely publicised stories about a 50% to 100% rise in infant mortality in the Harrisburg area after the accident, and predictions of 4000 to 8000 extra cancer deaths in the following decades. In fact the actual infant mortality statistics showed no change associated with the accident, and the radioactivity released is unlikely to cause even one cancer death. The accident was a major financial disaster. It took more than ten years to remove the damaged reactor core at a cost of nearly a billion dollars.

4.4 Chernobyl: The Disaster

Early in the morning of 27 April 1986 the radiation monitors around the nuclear reactor at Forsmark in Sweden registered a sudden increase, and the reactor was immediately shut down. Investigation showed that the reactor had been functioning normally, and that there had been no abnormal release of radioactivity.

Gradually, over the next few days, it became clear that there had been a great nuclear disaster deep inside the USSR, and that a radioactive cloud was drifting over Europe. Alarmist reports circulated about contaminated food and about the likelihood that thousands of people would die from cancer. The opponents of nuclear power were delighted: surely this would give the *coup de grace* to the hated nuclear power programme. Governments, some already hesitating about ordering nuclear reactors, immediately came under heavy pressure to abandon their plans.

The accident at Chernobyl was, without any doubt, a major disaster. Indeed in a lecture in Oxford, Zhores Medvedev, the Russian dissident and onetime Head of the Institute of Radiological Biology in Obninsk, said that it was one of the main factors leading to the collapse of the Soviet Union. A short time before Chernobyl, Mr Gorbachev's energy policy for the Soviet Union put increasing reliance on nuclear power. After Chernobyl, local anti-nuclear pressure stopped nearly all further construction and thus gravely exacerbated the economic difficulties that already faced him.

It is thus vital for the future of nuclear power to understand just what happened at Chernobyl, and whether such accidents are an inevitable feature of nuclear power stations. A meeting of experts was organised by the International Atomic Energy Agency and held in Vienna in August 1986, and this provided the first opportunity for Russian scientists to tell the world what had happened. Since that meeting there have been many others, so that we now know in considerable detail what took place on that fateful night, and during the following days.

The four reactors at the Chernobyl nuclear power station near Kiev are of the pressurised boiling water type. The uranium fuel is surrounded by graphite to slow down the neutrons so that they are able to cause more fissions. Water at high pressure flows between the fuel rods and is converted to steam by the heat of the reaction. This steam drives a turbine and produces electricity.

During the normal operation of the reactor at constant power the average number of neutrons from each uranium nucleus undergoing fission that causes another fission (called the multiplication factor) must be kept to exactly one. On the average, about two and a half neutrons are produced from each fission, so the balance of one and a half must either be absorbed by other materials in the reactor or escape through the walls. If the number of neutrons causing further fissions exceeds one the power level increases while if the number is less than one the power level decreases. It is therefore very important for the control of the reactor to know how rapidly this happens, so that the appropriate action can be taken in the event of a rise or fall in the power level. The time taken by a neutron from one fission to cause another one is extremely small, but fortunately there are just a few of them, less than one per cent, that are emitted from the fission fragments about a minute later. These are called the delayed neutrons.

If therefore we operate the reactor so that without these delayed neutrons the multiplication factor is less than one then it can easily be controlled, as we have a time of the order of a minute to make any necessary adjustments. If the multiplication factor exceeds one without the delayed neutrons the power of the reactor rises extremely quickly: this situation is highly dangerous and is referred to as prompt critical. It is thus vitally important in the operation of a reactor to ensure that this never happens.

The multiplication factor of a reactor is controlled by a series of rods made of neutron-absorbing material that can be moved in or out of the reactor. If there is any danger, these rods can be moved in rapidly, thus stopping the nuclear reaction.

A characteristic of the Chernobyl type of reactor is that if for any reason the core is heated then it automatically corrects itself when it is running at full power, thus ensuring stability. However, if it is running at low power it behaves in the opposite way and gets still hotter, so that it is thermally unstable. This by itself is not dangerous since it can be corrected by control equipment and running at a low power can be avoided.

Another feature of this type of reactor is that the neutrons are slowed down by the cooling water as well as by the graphite. This water also absorbs neutrons and so if the amount of water is reduced, for example by vaporisation, the neutron absorption is also reduced and this would increase the multiplication factor and so destabilise the reactor. These two sources of instability in the design of the Chernobyl reactor contributed to the disaster. Such a design would be unacceptable in Western countries on safety grounds.

The accident at Chernobyl did not happen during normal operation. It happened during an experiment that was made just before a routine service. The operators wanted to know whether the electricity generating equipment would continue to supply electricity for a short time when no longer supplied with steam. It was thus an electro-technical experiment, and the minds of the operators were on the electricity-generating equipment and not on the reactor. The experiment was directed by an electrical engineer, not by a nuclear scientist. They thought that the experiment would be spoiled if the reactor cut out so they disconnected the emergency core cooling circuit. Because of the instabilities mentioned, operation of the reactor at low power was forbidden, so they reduced the power to somewhat above that level. However the control rods were switched to the power reduction mode, and so the power level fell lower than planned. The operator then tried to increase the power level by removing the control rods still further, thus putting the reactor into a dangerously unstable condition. Additional cooling pumps were then switched on, thus increasing the amount of water in the reactor, and thence the absorption of neutrons. To maintain power, the control rods were still further removed. The operator then decided that all was now ready for the experiment, and to establish the condition of the reactor the computer printed out the distribution of power density and the positions of all the control rods. This showed that the reactor was in a highly unsafe condition forbidden by the rules of operation, and that it should then have been shut down immediately. Instead of doing this, the operator switched off the emergency safety mechanism and prepared to begin the experiment. Years of trouble-free operation had lulled him into a false state of confidence.

By then the reactor was in such an unstable condition that the power level began to rise. The operator saw this and ordered an immediate shutdown, but it was too late. The control rods could not move in fast enough to stop the rise of power. The reactor went prompt critical and in a few seconds the power level reached a hundred times full power. The fuel elements became extremely hot and burst open, vaporising the surrounding water.

The steam pressure blew off the top of the reactor and sent a shower of blazing graphite into the air. The graphite left in the core caught fire and burned fiercely for several days, sending a plume of radioactive smoke high into the atmosphere, whence it was blown over northern Europe. It is estimated that the radioactivity released amounted to 1.85×10^{18} Bq.

Fire-fighting teams raced to the reactor and with great courage tried to contain the damage. Thousands of tons of boron, clay and lead were dropped by helicopter into the blazing reactor and eventually the fire was put out. They and the reactor operators were exposed to very high levels of radiation and thirty-one of them died; subsequent deaths have raised this figure to 42. In addition, 237 more received high but not lethal doses. A huge rescue operation was organised in the surrounding area. The forty-five thousand inhabitants of the nearby town of Pripyat were evacuated to a safe distance, and tens of thousands of cattle were also removed. The radiation casualties were rushed to hospital and given the best available treatment; 207 of the 237 are now fully recovered, but have a higher chance of developing cancer in later life.

As the news of the disaster spread there was great public alarm, increased by the lack of reliable information. Governments had no experience of coping with such a catastrophe. The alarm spread to other countries as the radioactive cloud drifted over northern Europe. In many countries people were advised not to eat meat or fresh vegetables, as they had been contaminated by the rain from the radioactive cloud. In the following weeks accurate measurements of radioactivity were made all over Europe, and it is now possible to estimate the health hazards due to Chernobyl. In the immediate vicinity of the reactor the ground is heavily contaminated, and it will not be possible to use it until the radioactive material has been removed. Further away from the reactor, the distribution of radioactivity on the ground is very irregular, depending on the weather conditions at the time. Some areas are quite heavily contaminated but others, nearer the reactor, are not so badly affected.

To assess the damage caused by this radioactivity two questions must be answered. Firstly, how seriously is the surrounding region contaminated, and will it ever be safe for the inhabitants to return? Secondly, what will be the effects of the much smaller doses received throughout Europe and elsewhere?

It is easy to detect the radioactive decay of a single nucleus, so exceedingly minute amounts of radioactivity can be measured. It is therefore already well-known that practically everything is radioactive to some extent, including our own bodies. We are continuously irradiated by the cosmic radiation from outside the earth, and by the radioactivity in the ground. This constitutes the natural background of radiation, and provides a standard with which to assess the likely effects of any additional radiation due to the nuclear industry, to medical irradiation and to accidents like Chernobyl. Furthermore, the natural background varies very markedly from place to place, depending on the geology of the region. In Cornwall, for example, the natural background radiation is over twice the average for Britain, due to the granitic rocks that contain uranium.

Measurements of the radiation levels around Chernobyl show an excess over the natural level attributable to the accident. Detailed maps have been made, and show a very irregular distribution, impossible to summarise concisely. All that can be done is to give a few numbers to indicate the general levels found. All the following numbers are for levels of radiation expressed in terms of the unit microsieverts per hour.

The highest readings were of course near the reactor itself, being 30–50 at 300 metres. Further away, in the town of Pripyat, they were 0.5 to 0.9, and at the edge of the 30 km zone around the reactor, 0.7. In an administration building 18 km from the reactor, 0.2. For comparison, the Swedish scientists who made these measurements in September 1990 found 1.9 on the plane to Moscow. The average British exposure from natural radiation is 0.25, and that in Cornwall 0.85. From such figures it can be concluded that apart from the immediate vicinity of the reactor, the radiation levels are similar to those in Cornwall, or less.

This picture is confirmed by observations of the flora and fauna. Doreen Stoneham from the Oxford Research Laboratory for Archaeology visited the area in 1990 and found deserted gardens flourishing, with fruit trees laden with apricots, cherries, apples and greengages, and an abundance of flowers including rosebay, willow herb, wild chicory, thyme, cow parsley and verbascum. Birds were nesting in the chimneys of abandoned houses and in one nest there were two young storks. The only obvious damage was to the older pine trees.

Pripyat was derelict, with creepers growing over walls and pavements. Inside buildings the doses averaged 0.25, but outside they were much higher: 3–8 on the grass, 12 on exposed brick and 110 on another grass verge. She concluded that the contamination of the bricks is a difficult problem, but the biggest obstacle to re-occupation is the deterioration of the buildings themselves after years of neglect.

The health of the million or so people still living in 2700 settlements in the contaminated area was studied by 200 international experts from 22 countries co-ordinated by the International Atomic Energy Agency and the results were published in 1991 in an 800-page technical report. The project was led by the Director of the Hiroshima Radiation Effects Foundation and included members of the World Health Organisation, the International Labour Organisation, the UN Scientific Committee on the Effects of Atomic Radiation and three other independent organisations. It is thus difficult to accept the claims of certain environmental groups that their report was a political whitewash and that they were duped by the Soviet authorities. The report found no health disorders directly attributable to radiation exposure, and in particular no indications of increased mortality rates, still births and incidence of cancers, including leukaemia. This was not surprising, because radiation-induced cancers often take years to develop. To establish the presence of any adverse health effect, it is necessary to compare the observed incidence with that expected in the absence of the accident, making allowance for any changes in lifestyles and also in the diagnostic techniques. Since many people were irradiated, it was expected that there might be a detectable increase of some cancers after about seven years. In particular, some population groups received thyroid doses around 1 Sv, and several hundred cases of childhood thyroid cancer due to the radioactive iodine 131 were indeed found during 1991–4 in Belarus and the Ukraine, rather sooner than expected. The rates were so greatly in excess of those occurring before the accident that they cannot all be ascribed to improved diagnostic techniques. It has been suggested that they may be due to the potassium iodide tablets that were given to block the effects of the radioactive iodine since the doses given to some children in Belarus and the Ukraine in 1986 were far too high. This suggestion is supported by the statistics for Russia where no preventive iodine was given and just two cases of childhood cancer occurred. The recorded increase coincided with the introduction of ultrasonic screening, which often reveals cancerous nodules that may not develop into observable cancers for several years. It should be noted that thyroid cancer can be treated easily if it is diagnosed early. There was no evidence of any other effects, particularly of childhood leukaemia, congenital abnormalities or any other radiation disease either in the contaminated regions or in Western Europe.

It has also been reported that there have been 6000 deaths among the workers engaged on the cleanup operation, but this is just about the number that would have been expected from other causes in the very large number of people involved (some 800,000) over the period in question. Many of these workers suffered from radiation sickness, but nevertheless the death rate from all causes is similar to that in the general population.

The psychological effects of the accident were widespread and severe. Nuclear radiations are unseen and the information given to the public was not enough to allay their fears that they had been subjected to dangerous radiation. They had been told that such an accident was impossible and yet it happened. Thousands of people were uprooted from their homes and told to avoid certain familiar foods. The official reactions, understandably enough, were not always consistent. It was the time of glasnost and perestroika, and Chernobyl came to symbolise all that was wrong with the old system. In this situation there was a tendency to attribute every illness to the effects of Chernobyl, whether or not it is of a type that can be caused by radiation. To deal with this situation, every sufferer from leukaemia was officially classified as a "victim of Chernobyl", and is entitled to compensation, although there is no evidence that the illness is actually due to Chernobyl. Groups of children who have been sent to recuperate in other countries have been found to suffer from malnutrition, but not from the effects of radiation.

This conclusion is supported by the results of Professor Kellerer from Germany, who found that the people in the Chernobyl area thought that their medical conditions were the result of radiation, and did not understand "that the undoubted deterioration was the result of changed lifestyles and an enormous degree of anxiety, a sort of self-amplifying problem. People believe they are surrounded by problems and they do not drink local milk, abandon their cattle, rear no poultry and keep children indoors. This whole thing leads to a breakdown of lifestyle. So quite understandably you have indeed increased morbidity".

Undoubtedly there was much suffering due to the evacuation itself, and even more due to the very natural fears of the effects of radiation, not helped by the media. Indeed, the international project commented: "The psychological problems were wholly disproportionate to the biological significance of the radioactive contamination. The consequence of the accident are inextricably linked with the many socio-economic and political developments that were occurring in the USSR. A large proportion of the population have serious concerns. The vast majority of adults examined in both contaminated and control settlements either believed or suspected they had illness due to radiation".

This was underlined by a report from four Swedish scientists who visited the area in 1990. They were distressed to find evidence of widespread ill-health, but noted that this was not induced by radiation. They believe that "many people are exploiting the biggest consequences of the Chernobyl accident — the radiophobia — to further their own aims". Before the disaster there was chronic malnutrition, and now people are afraid to eat or go outdoors.

Soviet experts are well aware of this, but their reassurances are undermined by local officials with a political interest in maximising the consequences of the disaster in order to obtain more financial assistance. The Swedish scientists interviewed Professor Guskowa, a member of the Soviet Academy of Science and noted: "When we took our leave of Angelina Guskowa we said goodbye to a very pessimistic woman who all her professional life has been engaged in work at home and abroad to cure people suffering from the effects of radiation, to spread information about these effects, and who has endeavoured to estimate the risks of radiation. It was obvious that she was disappointed in how the egotistical interests of politicians and suchlike who, totally without support for such a claim, pose as experts, have managed to sabotage everything she has worked for".

What of the effects of the radioactive contamination of Europe as a whole? Here again the deposition was extremely patchy, depending on the winds and rain. Saltzburg even received more than Kiev. The additional radiation received by people in Europe range from the insignificant to around the level of the natural background, so it is not easy to estimate the effects on health. The effects of large radiation doses are well-known, but almost nothing is known for certain about the effects of very small doses. It could be that the body is able to repair any damage due to very small doses. There is even some evidence that very small doses are beneficial to health. In Cornwall, for example, where the natural radiation background is higher, the incidence of leukaemia is rather less than in the rest of Britain. Nevertheless, in spite of this, it is usual to assume that the effect of radiation is strictly proportional to the dose. This is almost certainly a very pessimistic assumption.

Making the assumption of proportionality, it is easy to obtain figures like the 40,000 deaths worldwide over the next 50 years quoted in the media. A more realistic estimate made by the National Radiological Protection Board is of the order of a thousand. This would be statistically undetectable among the 500 million people expected to die of cancer in any case in the same period.

Once again the natural background provides a standard for judgement. The extra doses due to Chernobyl are quite small compared with other increases we accept without hesitation, such as those on airplane flights, on mountains (due to the cosmic radiation, which becomes more intense with increasing height) and due to the variations from one place to another. In Britain, the average extra dose was about thirty times less than the natural background. There has been discussion about whether more people should be evacuated from the Chernobyl neighbourhood, but none about the evacuation of Cornwall, which has a natural radiation level of about 8 mSv/yr, with peaks in some regions of over 100 mS/yr. It is estimated that the residual effects of Chernobyl are as hazardous as remaining indoors for one minute longer each day, thus increasing the radon dose. A comparison between the lifetime doses due to Chernobyl and the natural backgrounds in three European countries is shown in Fig. 4.1.

There were even some reports of radiation effects beyond Europe. A much-publicised news item reported a strong correlation between mortality rates

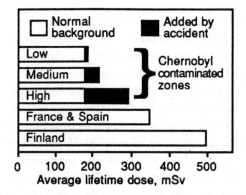

Fig. 4.1. The 70-year lifetime doses due to Chernobyl compared with those due to the natural background around Chernobyl and in several European countries (*Nuclear Issues*, April 1996).

and the radiation levels from Chernobyl dust in the USA. This looked very convincing, until one reflected that if such minuscule amounts of dust could have such pronounced effects in the USA then the much larger (though still very small) amounts received in Europe would have spectacular effects, which were not present. Examination of the basic data showed that the alleged correlation was spurious. By that time the media had lost interest in the story and had gone on to the next radiation scare.

After Chernobyl, several countries, such as Sweden and Switzerland, voted to phase out nuclear power as soon as practicable. But when they looked into the alternatives they realised that they were even less attractive. Coal is seriously polluting, and is contributing to acid rain and to the greenhouse effect. Oil will become more expensive, and it is politically undesirable to rely on it. Hydroelectric power is already used to the practicable limit in most European countries, and the renewables such as wind and solar are not credible as a large-scale source. If these sources are excluded, and nuclear power is phased out, the price of electricity will rise, with serious effects on welfare and employment. It has been estimated that it would cost Sweden $1B/yr for ten years to replace the nuclear power stations. So the resolutions to phase out nuclear power are being quietly forgotten. In Russia the nuclear power programme is continuing, for there is no other way to meet the expanding demand for electricity.

Another panic reaction of the Swedish Government to Chernobyl was the burying of thousands of reindeer carcasses in Lapland because they were slightly contaminated by radioactivity, thus depriving the Lapps of meat and the income from surplus sales. Swedish radiation physicists have criticised the radiation limit of 300 Bq/hr for reindeer meat (equivalent to 3000 Bq/yr for 10 kg of meat) as totally irresponsible. In the same units, the natural background is 38,000 Bq/yr and the average dose from radon in homes is 120,000 Bq/yr. The head of the health physics department of the Riso National Laboratory in Denmark said that the consumption of 70 kg of meat containing 150,000 Bq/kg would incur a risk comparable with that of smoking one cigarette a week. In Germany, about $300M were spent decontaminating cattle feed additives, yet the average 50-year exposure of the population is only about 1% of the natural background. Worldwide, the use of phosphate fertilisers has added five times as much surface radioactive contamination as Chernobyl.

Prince Philip has remarked that "over the last hundred years acid rain has done a thousand times more damage to the environment than the one accident at Chernobyl". The most enduring part of the tragedy of Chernobyl is not the result of exposure to radiation but the social and psychological stresses and the anxiety this causes. Taken together, these factors affect the health of the people far more than the radiation exposures.

4.5   The Lessons of Chernobyl

The whole story of Chernobyl is one of tragic and misguided muddle that had brought death to some people and suffering to many hundreds of thousands of others. It is important to examine it carefully, with due regard for the facts, to see what lessons can be drawn for the future.

The nuclear physicists and engineers have developed a new source of power that has many advantages over the alternatives as regards availability, capacity, safety, cost and effects on the environment. But like most modern technologies, it is complicated and has to be handled with due care. This is not confined to nuclear power; it applies to all industrial processes and modes of transport. The choice is ours, both individually and as members of society. We do not entrust a car, let alone an airliner, to one who is not properly trained; if we are to do so we would be asking for trouble. In the case of new technologies, it is the duty of governments to lay down and enforce strict rules of operation. Indeed, nowadays, when the effects of a major disaster are felt worldwide, it is arguably the duty of the United Nations to do this.

At this point the duty of the scientists and engineers cease, and that of the politicians begin. It is now clear that the politicians in the Soviet Union failed in this duty. The reactors were built hurriedly for the dual purpose of producing weapons grade plutonium and providing power for civil use. To achieve this, a design was adopted that would never have been accepted in the West. It was basically unsafe, in the sense that when operating at low power it was thermally unstable; that is, a small extra power increase reinforced the increase instead of producing a cooling that would tend to reduce the power to the desired level. To prevent a dangerous situation developing, strict operating instructions were laid down.

Unfortunately people do not always obey instructions, and on the night of the disaster the reactor was run at low power so that the operators could conduct a certain experiment. In the interests of safety, the reactor was designed to shut down automatically if it was operated in the unstable region. However the operators realised that this might spoil their experiment, so they switched off the safety devices, and disaster followed. There was no one in the control room who understood the risks that they were taking by trying to make this experiment.

All this could never have happened if the design and construction of reactors had been carried out under the control of an international regulating body.

When the disaster occurred, the scientists and technicians around the reactor, aided by fire-fighters, fought with extreme bravery to control the reactor, and many paid with their lives. The politicians reacted by imposing secrecy until the accident became known internationally and they had to admit that it occurred. The evacuation of people was probably essential in the circumstances, but not enough was done to give the people accurate knowledge concerning the hazards associated with the doses they received.

Local politicians reacted by using the disaster to advance their own political ambitions. The politicians in other countries reacted with panic and voted to phase out nuclear power without a proper study of the implications. The people most affected, who lived in the area around the reactor, were not given accurate information about radiation hazards, and so they were terrified by the inflated stories abounding in the press and many of them suffered severe psychological problems. They became apathetic and feared to eat readily available food. In the end the danger to health from this were far larger than the dangers attributable to radiation.

Soon after the accident, a massive clean-up operation was started, and this has continued with international assistance. Extensive measurements have been made of radiation levels to determine when it is safe for the population to return. A major problem remains the healing of the deep psychological wounds.

And what of the Chernobyl-type reactors, not only in what was the USSR but also in Eastern Europe? Many if not most of them are seriously inadequate by Western standards. Should not they be shut down at once? This extreme measure is unnecessary, and would cause widespread hardship and serious disruption of the economy. The Chernobyl disaster occurred as a result not only of design faults but also of a whole series of gross operating errors, including switching off the safety devices. With responsible operating procedures, they can be operated safely. Nevertheless it remains urgent to modify them to bring them up to international standards, and this is now being done with the help of the International Atomic Energy Agency.

References

Chernobyl: Ten Years On. Radiological and Health Impacts, NEA Committee on Radiation Protection and Public Health, Nuclear Energy Agency, 1995.

Gale, R., *The Health Impact of Chernobyl, Nuclear Europe* 5/1987.

Medvedev, Z., *The Legacy of Chernobyl*, Blackwell, 1990.

Shcherbak, Y. M., *Ten Years of the Chernobyl Era, Scientific American* April, 1996.

5

CHOOSING THE
BEST ENERGY SOURCES

5.1 Introduction

In the last four chapters the various possible energy sources have been described, and some of their advantages and disadvantages outlined. To decide the best energy policy, all these sources must be evaluated and compared, as far as possible using objective criteria. There are five criteria that can be used, namely capacity, cost, safety, reliability and effects on the environment. In some respects this can be done globally, and in others it must be done for each country, taking into account its needs and natural resources. To ensure objectivity, the comparisons must be made numerically wherever possible. This can be done fairly well for the capacity, the cost and the reliability, and to some extent for safety, but hardly at all when we consider the effects on the environment.

This task would be difficult enough if we had all the necessary data, and the situation was static. Unfortunately this is not so. The data are inevitably insufficient, and the energy needs are continually changing. It is not practicable to say that we must postpone decisions until more data are available; decisions have to be made now. It takes many years to build a new power station and if we wait until we are sure that it is needed, then it will probably be too late.

For the same reason the decisions must be made considering existing technologies that have already been tested and have shown their worth. There are many bright ideas about new energy sources, or new ways of improving old ones. They may or may not prove viable in the long run, but we cannot wait until they are properly tested. When they are proved successful, then they can be considered as possible new sources in competition with those already existing.

Rapid technological change may not only provide new energy sources; it may radically alter the energy demand in ways we cannot foresee. When we estimate future needs, or any other variable, we extrapolate the measurements made over the last few years. This is the best we can do, but it may easily be seriously wrong. To take an extreme example, we may imagine anxious discussions among stone age men about the future supply of flints. We know that their worries were unfounded; technology moved forward and quite different raw materials were in demand. In recent years we have seen a range of plastics used to replace many products previously made of metal. The demand for steel is less than might have been predicted, and many steelworks have been forced to close. These difficulties should not stop us trying to see into the future, but they should induce some caution.

The following sections assess the major energy sources according to the suggested criteria of capacity, cost, safety and reliability. The effects on the environment are considered in the next chapter. Although they are separated for convenience of discussion, these criteria are interconnected. Thus, for example, if we want to increase the safety of a particular method of energy generation, or to reduce the harmful effects on the environment, this will inevitably increase the costs and thus the desirability of that particular energy source. Some conclusions are given in Sec. 5.6.

5.2   Capacity

Broadly speaking, energy sources may be divided into three classes according to their ability to provide the energy needed by an industrialised country. In the first class come those that could provide it all, and these are coal, oil and its associated natural gas, and nuclear. Fusion is also potentially in this class, but as it has not yet been developed to the stage of a power reactor it is not further considered here. Wood, as we saw in Chap. 2, is not suitable as a large-scale energy source.

In the second class comes hydroelectricity, which depends on the availability of suitable rivers. In mountainous countries like Norway and Switzerland it can supply a large fraction of the energy needed, but in flat countries like Denmark none at all. All over the world, hydro can contribute at most about 10% of the energy needs. This is very important, but it is not sufficient to solve the energy crisis.

In the third class comes geothermal energy and the other renewable energies that for the foreseeable future are unlikely to provide more than a few per cent of the world energy needs.

The capacities of the major energy sources will now be considered in more detail.

The coalfields of Britain provided the energy that made the industrial revolution possible. Coal is still of vital importance and is likely to remain so for the foreseeable future. Coal has many advantages as a source of energy. It is relatively easy to mine, contains energy in a safe and transportable form, and it is easy to release the energy when needed in small or large quantities. This is why it rapidly replaced wood for domestic heating and industrial purposes, and made steamships possible. Its derivatives, coke and gas, found applications for heating and lighting.

There are vast coal resources and reserves mainly in the former USSR, USA and China, and to a lesser extent in Australia, Poland, South Africa, Germany and the UK, as shown in Table 5.1. The total resources amount to about 10^{12} tons of coal equivalent. This unit is used to take into account the different calorific values of various types of coal. Only about 6% of this is estimated to be economically recoverable at present. Thus if the price of coal rises, which seems very likely, so will the world reserves.

Table 5.1. World Coal Reserves at the end of 1997 (in 10^9 tons).

	Anthracite and Bituminous	Sub-bituminous and Lignite	Reserves/ Annual Production
USA	106	134	244
South + Central America	6	5	231
Germany	24	43	301
Poland	29	13	209
UK	2	1	52
Former Soviet Union	104	137	> 500
S. Africa	55	—	255
Australia	45	46	327
China	62	52	82
India	68	2	212
TOTAL*	519	512	219

*Includes countries with smaller reserves omitted from the Table.
Sources: BP Statistical review of World Energy, June 1998.

The present world coal production is about 2.6×10^9 t.c.e. and it is estimated that this could be tripled by the year 2020. This implies an annual growth rate of 2.6%, which is very similar to that achieved in the last century. It is notable that only about 8% of coal is exported, most coal-producing countries being primarily concerned with satisfying their own needs. Since the major coal producers are in the northern hemisphere, this implies that a substantial increase in the production of coal for export is essential if coal is to contribute to the energy needs of all countries.

The last column in Table 5.1 shows the reserves divided by the annual coal production. These figures show that enough coal is available with existing technology to last about 200 years. There is thus no immediate worry about resources and reserves; the problem is to increase the rate of coal production to contribute as massively as possible to world energy needs.

There are however limits to the rate of increase of coal production set by the high cost of new mining facilities and the time it takes to build them. A report in 1993 on the British coal mining industry said that it is uneconomic to open new pits, and that at the present rate of extraction the coal in existing pits would last only 21 years. Coal mining is a hazardous occupation and it is increasingly difficult to find men willing to accept the risks and discomforts. Another difficulty is pollution, which will be considered in more detail in Chap. 6.

During the last century, and increasingly in the present century, the discovery of vast oilfields in the Middle East, Russia, North America and more recently South America, Alaska and the North Sea has provided an abundant fuel with many advantages over coal. It is much easier to extract from the ground; in many cases if a well is drilled the oil gushes out. It is easier to transport over large distances by pipeline or tanker and it has a high calorific value. The primary oil varies in quality and it is progressively distilled to give a wide range of fuels from petroleum to heavy oils and finally tars.

This abundant and convenient fuel has progressively replaced coal during the present century, particularly in heavy industry, transport (ships and trains), power stations and domestic heating. In what was West Germany, for example, the portion of the energy market using oil or natural gas rose from 5% in 1950 to 69% in 1978, while that of coal fell from 88% to 27% in the same period. Oil has also made possible the development of cars and airplanes, which cannot be powered directly by coal. Thus in a relatively short time a large part of the world's economy has become heavily dependent on oil.

It has proved so easy to extract large quantities of oil that at least until 1973 its price compared favourably with other fuels. As a result, the oil reserves are being more rapidly being used up than new oilfields are being found. With the addition of deep off-shore and polar regions, the total world resources amount to about 330 GT. Most of these resources are quite accessible and it is estimated that more than half can be produced at less than 1976 selling prices, and a third at about $5 per barrel. The 1977 level of consumption is about 3 GT per year, so even if this rate remains the same, the oil will be exhausted in about 100 years. However the demand for oil could rise considerably, and in response the oil production could rise to 4 or 5 GT per year, thus reducing the life of the oil reserves to about 60 years.

As in the case of coal, the increasing costs of developing less accessible oil fields is offset to some extent by technological improvements and enhanced recovery rates, and the rising price of oil justifies the development of the less profitable fields.

Only a relatively small proportion of the world's oil resources is economically recoverable and much research is in progress to increase the recovery rate. As an indication of the size of the problem, it has been estimated that 450 billion barrels of oil have been discovered in the United States and of this 115 billion barrels have already been extracted and 34 billion barrels remain to be extracted, while the remaining 301 billion barrels cannot be extracted at the present time. Several processes for increasing the extraction percentage have been suggested, including pumping steam, chemicals or gases into the rock to drive out more oil. The best method to use depends on the particular circumstances. Often these techniques are expensive, so their practicability is restricted on economic grounds.

Some rocks contain a tar-like substance called kerogen that releases a viscous oil when heated to about 500°C. This oil can be refined to give petroleum. About two tons of rock must be processed to give a barrel of oil, so the disposal of the exhausted rock is a severe environmental problem. The cost of petroleum from this oil shale is substantially higher than that from oil, and pollution control could increase it still further. Although the deposits are huge, the equivalent of 600 billion barrels in the American West alone, it is thus unlikely that this process will make a significant contribution to world needs. Petroleum can also be produced from coal, but not economically. Because of these uncertainties, petroleum from such sources are not included in the Table. It should also be noted that the extraction of petroleum from tar sand, oil shales and coal itself requires energy, and is also a considerable source of atmospheric pollution. The cost of petroleum from such unconventional sources is estimated to be at least $30 to $50 per barrel, which restricts it to specialised uses such as transport and chemical manufacture.

The conclusion of these studies is unmistakable: within a lifetime the massive contribution of oil to the world's increasing energy needs will be falling steadily. Inevitably the price will rise, and more and more applications will be rendered uneconomic. The world will be forced to reduce its reliance on oil.

Natural gas, often found in association with oil, is one of the most convenient of all energy sources. It is easy to extract from the earth, has a high calorific value, is relatively non-polluting, and can easily be piped over large distances. Natural gas is found in many parts of the world, particularly the former USSR, East Europe and some OPEC, and to a lesser extent Western Europe.

One of the main disadvantages of natural gas is that it can only be transported over large distances by tankers after liquefaction. Often this is not economic, so in many cases where oil and gas occur together, the gas is burnt at the well-head. A higher utilisation of gas would be stimulated if the price were to rise to make its transport economically worthwhile.

The annual energy production from natural gas is about 50 EJ, and reserves are estimated at 2,500 EJ, with undiscovered resources about 50,000 EJ. It is likely that the annual production will double or triple by the year 2000 and thereafter decline. The resources are large enough to maintain production at double the current rate for at least fifty years. Thus natural gas makes an important though not dominant contribution to world energy needs.

The increasing demand for natural gas has simulated the development of processes for the gasification of coal. This can be done on a commercial scale, but the cost of the gas is several times that of natural gas. This is still economically useful in areas far away from natural gas fields, and may prove to be the only practicable way to use sulphurous coals in remote areas. The gas may either be transported to where it is needed or burnt on the spot to drive steam turbines to generate electricity. The efficiency of these systems has been increasing steadily with the development of new alloys able to withstand high temperatures.

There are now 442 nuclear power reactors in operation in many countries worldwide with a total capacity of over 350,000 MW, and in 1996 they supplied 2300 TWh, about 20% of the world's electricity. The number in each country

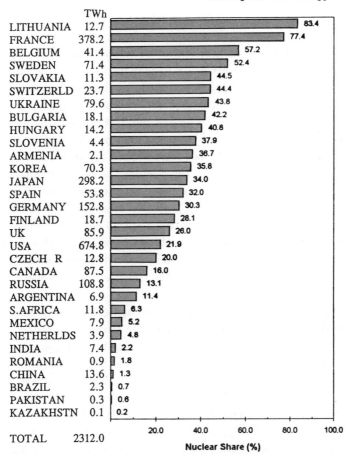

	TWh	
LITHUANIA	12.7	83.4
FRANCE	378.2	77.4
BELGIUM	41.4	57.2
SWEDEN	71.4	52.4
SLOVAKIA	11.3	44.5
SWITZERLD	23.7	44.4
UKRAINE	79.6	43.8
BULGARIA	18.1	42.2
HUNGARY	14.2	40.8
SLOVENIA	4.4	37.9
ARMENIA	2.1	36.7
KOREA	70.3	35.8
JAPAN	298.2	34.0
SPAIN	53.8	32.0
GERMANY	152.8	30.3
FINLAND	18.7	28.1
UK	85.9	26.0
USA	674.8	21.9
CZECH R	12.8	20.0
CANADA	87.5	16.0
RUSSIA	108.8	13.1
ARGENTINA	6.9	11.4
S.AFRICA	11.8	6.3
MEXICO	7.9	5.2
NETHERLDS	3.9	4.8
INDIA	7.4	2.2
ROMANIA	0.9	1.8
CHINA	13.6	1.3
BRAZIL	2.3	0.7
PAKISTAN	0.3	0.6
KAZAKHSTN	0.1	0.2
TOTAL	2312.0	

Nuclear Share (%)

Fig. 5.1. Nuclear energy generation in 1996. (Reproduced from "Energy, Electricity and Nuclear Power Estimates for the periods up to 2013" (July 1997 Edition) by kind permission of the International Atomic Energy Agency, Vienna).

depends on the availability of other energy sources, especially of coal and oil. In France, for example, a highly industrialised country with practically no indigenous coal or oil, nuclear power is now responsible for over 75% of the electricity generated. In Western Europe, nuclear power has overtaken coal, and supplies about 50% of the electricity. In many other countries a very substantial amount of the electricity is generated by nuclear reactors, as shown in Fig. 5.1. There is thus no doubt that nuclear power has the capacity to supply a large fraction of world energy needs.

5.3 Cost

It is not as easy as might at first appear to estimate the cost of generating energy in various ways. It is necessary to take into account the initial cost of building the power station and the associated distribution facilities, the cost of fuel and maintenance while it is running and the cost of decommissioning when it has reached the end of its life. The first and last of these are generally the largest, and their contribution to the cost of energy is divided by the number of years of operation, and this is not known when the power station is built. The longer the power station lasts, the cheaper the energy it produces. It is therefore essential to have experience based on the operation of several full-scale power stations for a number of years, for otherwise it is not possible to take into account various eventualities that were not envisaged when the power stations were designed. Ideally it is better to have information from several different types of power stations operated in different countries to take account of differing costs of raw materials, and the construction and operation of the power stations.

Some types of power stations, particularly nuclear, are expensive to build but relatively cheap to run, so the cost of the electricity produced depends strongly on the rate of interest required on the initial capital expenditure. Likewise, the number of years of power-producing life greatly affects the cost per unit. Thus, for example, at the time of privatisation, nuclear power was required to recover the fixed costs in twenty years (instead of the predicted lifetime of forty years) and earn interest at 10%. The effect was to add 2.49 p/kWh to the estimated cost of 3.23 p/kWh for Hinkley Point C. With decommissioning and fuel reprocessing adding 0.54 p/kWh, this gives a total of 6.26 p/kWh. Apart from the somewhat arbitrary choice of interest rate, the main uncertainty is the lifetime of the reactor. This can only be known with confidence as a result of actual experience. Until we have run a plant for years we do not know how many unexpected faults may develop. Things wear out, corrode and break, and repairs increase the cost and reduce the power output. This uncertainty is particularly serious for new designs, so any estimates of the lifetime not supported by actual running experience should be treated with great caution.

Some power stations take five or more years to build, and have a useful life of forty or fifty years, and in that time inflation and other social changes will drastically alter the value of the currency. The cost of the fuel may vary substantially during this time; as an indication of this, the cost of coal in recent years has varied by a factor of three and oil by a factor of five. The maturity of a particular technology is also an important factor. The first few power stations built using a new energy source are inevitably more costly that those built later on. Construction and operating experience leads to improvements in design and increased efficiency. There are also improvements due to scale: it is cheaper per unit to build several power stations of the same design and at the same place. The cost of decommissioning a nuclear reactor at the end of its life is difficult to estimate because it is so far into the future, and there may be little experience of the processes involved. The decision to close a power station depends on technical matters, but is also strongly affected by political and economic considerations, and by safety and environmental regulations. Since these change in an unpredictable way, it is practicably impossible to estimate reliably the life of power stations when it is first built. As an example, the Magnox nuclear power stations in Britain were designed to last twenty years, but in 1991 they were cleared for up to thirty years. It has been suggested that the operating lifetimes of the more advanced designs might be 45 or even sixty years. For the extra period the cost of nuclear electricity is then only the marginal cost of 2 p/kWh.

The running costs of power production are also strongly affected by regulations concerning safety and pollution. In some cases lengthy and expensive hearings are conducted before permission to build is granted, and restrictions on emissions may require the installation of costly purifying equipment. There may also be direct or hidden state subsidies to encourage a particular type of power station for political reasons. For all these reasons it is often extremely difficult to obtain a realistic figure for the cost of energy production. All such figures, including those given below, should be treated with much caution. Often one of the best indications of relative cost is given very simply by the choices actually made by Governments and industrial companies, who are likely to make detailed studies before making the very large investments needed. This still leaves open the possibility of mistakes and also one can challenge the justification for the safety and pollution regulations, which may be unduly influenced by political considerations.

It is also useful to know the time it takes for a power station to generate the energy used in its construction. If this is greater than its lifetime then no net energy is produced. This is important for some of the more speculative renewable energy sources.

With all these reservations, we can attempt to estimate the costs of various types of energy. Since the main concern is energy on a large scale it is simplest, and also facilitates comparisons, to estimate the cost of generating electricity for the main sources, coal, oil and nuclear. Due to inflation the relative costs are likely to be more reliable than the absolute costs when account is taken of the variations of the interest rates.

To show how cost estimates have changed over the years, it is instructive to give figures obtained at various times. The Report of the Central Electricity Generating Board for the year ending March 1981 gave the total average costs for electricity from the nuclear (Magnox), coal and oil power stations as 1.65, 1.85 and 2.62 p/kWh respectively. Very similar figures are given for the newer power stations, the Hinkley Point Advanced Gas Cooled Reactor (AGR) and the Drax coal-fired stations, which produce electricity at 1.45 and 1.87 p/kWh respectively. Comparable figures reported by Electricite de France are 1.5, 2.7 and 5.4 p/kWh for nuclear, coal and oil respectively.

Similar comparisons can be made for the operating costs on their own. A survey of 54 nuclear, 25 coal-fired and 18 oil-fired power plants of 400 MWe or over in the United States yielded average operating costs of 0.8 p/Kwh (nuclear) 1.0 p/kWh (coal) and 1.8 p/kWh (oil). In France the comparable figures are 0.8 p/kWh (pressurised water reactors), 1.3 p/kWh (coal) and 1.9 p/kWh (oil). Due to the rise in construction costs, the figures for later power stations are considerably higher. A survey in 1987 of the efficiency of energy generation in the USA showed that of the top twenty power stations sixteen were nuclear.

In 1990, the costs were 2.75 p/kWh for Scottish Nuclear, 2.86 for National Power, 3.03 for PowerGen and 4.31 for Nuclear Electric. The cost for the Magnox reactors is quoted at 4.8, for the AGR 5.5, for Sizewell 2.5 to 3.3 for 8% interest rate. The latter figure may be compared with 2.2 to 2.8 for a combined cycle gas turbine. Swedish reactors produce electricity for 1.7 p/kWh. In Spain, for the period January to October 1988, the total generation costs in pts/kWh were hydro 3.18, nuclear 6.94, anthracite 8.71, lignite 9.11 and imported coal 8.8. Some more comparative costs are given in Tables 5.2 and 5.3.

There are now many cases where nuclear electricity is exported from one country to another. In particular France, with its large nuclear programme, exports electricity to Britain, Italy and Switzerland. Britain can buy from France at two-thirds of the cost of its own electricity. The cheaper electricity in France benefits French industry by about £1B per year compared with British industry.

When a nuclear reactor has reached the end of its useful life it has to be decommissioned, and this is much more difficult and therefore costly than for

Table 5.2. Comparative costs of a kW of electricity from various sources in several countries.

Country	Year	Nuclear	Coal	Oil	Gas	Currency
Finland	1977–82	0.121	0.165	—	—	Fmk
Japan	1982	12	15	20	19	yen
Japan	1989	8	10	11	—	yen
France	1984	19.3	31.4	66.8	—	cent
France	1992	2.8–3	3.3–4	—	3.4–4.1	pence
Britain	1994	2.9	2.6–4.6	—	2.7–4.2	pence
Netherlands	2000	8.7	11.2	—	—	cent

Table 5.3. Comparative costs in p/kWh of electricity from nuclear and coal power stations in 1995.

	UK	FRG	Italy	Holland	Belgium	France
Nuclear	3.95–4.91	3.68	3.21	3.30	2.95	2.67
Coal	6.38	5.79	4.29	4.10	4.52	4.56

other types of power stations because the reactor core is highly radioactive. It is relatively easy to remove the fuel rods, which contain nearly all the radioactivity, and transport them to a reprocessing plant. Much of the external building and parts of the reactor are not radioactive and can simply be removed. This leaves the reactor core, and this can either be allowed to decay for many decades before dismantling or, more simply it could be sealed and buried under a mound of earth. This latter procedure would substantially reduce the cost of decommissioning. It has been estimated that the additional cost due to decommissioning is around 0.13 p/kWh for Magnox reactors, 0.1 for AGR's and 0.03 to 0.05 p/kWh for PWR's like that at Sizewell.

The relative costs of coal and nuclear power depend on a number of factors such as the proximity of coalfields, and detailed studies have shown that in some cases they are about the same and in others coal is about twice as costly as nuclear. The basic reason for the relative cost advantage of nuclear is that the energy is extremely concentrated. There is as much energy in a pound of uranium as in a thousand tons of coal. To some extent this advantage is reduced because nuclear power stations are so costly to build, but their running costs are less because such small amounts of fuel have to be transported to the power station.

Nuclear power costs have remained remarkably stable over the last two decades, but the substantial fall in the cost of coal has put it at a relative disadvantage. Nevertheless in 1992 a study by the OECD Nuclear Energy Agency and the International Atomic Energy Agency showed that for a range of countries the costs of generating electricity from coal, gas and nuclear are quite similar.

The costs of the various energy sources, and the available resources, are all inter-related. Thus if the price of one energy source rises it is likely to increase the demand for alternative energy sources, thus also increasing their prices. Conversely, a cheap form of energy will be increasingly used at the expense of other sources, which will then be obliged to lower their prices in order to survive, and may be rendered uneconomic. A price rise immediately increases the reserves because it renders economic processes that were previously uneconomic. Thus the sharp rise in oil prices in 1973 made economic many oilfields such as the North Sea and Prudhoe Bay in Alaska. High prices also encourage increased efficiency and the exploitation of new resources such as oil shale.

It should be emphasised once again that the relative costs of different energy sources depends on the interest rate, mainly because of the relatively high construction costs of nuclear power stations. If this is 5% per annum, nuclear power is the cheapest source for 13 out of 15 countries, the exceptions being regions with cheap coal (Western USA and Western Canada) or cheap gas (UK). At an interest rate of 10%, nuclear has an advantage in only four countries.

It is now increasingly recognised that in addition to the direct costs of producing energy there are many indirect costs, sometimes referred to as the social costs. These include all the effects on the environment due to emissions from power stations that pollute the air leading to acid rain, with deleterious effects on our health and indeed the whole biosphere. The land is polluted, also affecting our health, and the sea is polluted, harming the fish. The workers in the energy industries, especially coal miners, often have their health permanently affected, and this imposes social costs on their families and on society as a whole. The greenhouse effect could change the world's climate and inundate low-lying countries, with incalculable consequences. Sometimes included among the social costs are the Government subsidies to the energy industry. Apart from these, it is very difficult to quantify the social costs because they involve many personal assessments such as the value of a forest or of a species of fish.

Quite reasonably, it has been suggested that these social costs should be added to the price of the power produced. This would bring home to people what they are actually doing to the earth, on which we all depend, and would have the salutary effects of reducing demand and bringing about a new sense of social responsibility. The difficulty is to evaluate social costs and then to persuade the energy industry and the people to accept them. The proposal of

Table 5.4a. Estimated costs per unit of energy production attributable to environmental damage. Dollars (1990) per Giga-Joule.

	Coal	Oil	Gas
Mining	0.46		
Oil spills	0.85		
Atmospheric pollution			
Effect on man	4.51	3.60	1.50
Effect on animals	0.27	0.22	0.99
Effect on agriculture	0.58	0.46	0.19
Acid Rain			
Effect on aquatic ecology	0.16	0.13	0.05
Effect on forests	0.65	0.52	0.21
Effect on agriculture	0.29	0.23	0.10
Effect on buildings	0.79	0.63	0.26
Global warming			
Effect of climatic change	3.37	2.74	2.06
Effect of increased sea level	0.42	0.33	0.25
Total	11.50	9.71	4.71

Table 5.4b. Real cost factors.

Fuel	Coal	Oil	Gas
Market cost (a)	2.19	17.25	8.85
Real cost (b)	13.69	26.96	13.56
Cost factor (c = b/a)	6.26	1.56	1.53
Fraction of consumption (d)	0.60	0.20	0.20
Weighting factor (h = d × c)	3.75	0.31	0.31
Total factor		4.37	

Energia et Innovazione, 1988

a carbon tax, to reduce carbon dioxide emissions, has already aroused strong opposition. It is obviously difficult to quantify the cost of health and that of the environment. It has also been estimated that taking into account the social costs the cost of coal power could become four times that of nuclear. Some additional estimates are given in Table 5.4

In 1991, the Chairman of Nuclear Electric, John Collier, estimated the environmental costs of fossil fuels. He found that if a carbon tax of $10 per barrel of oil equivalent were imposed, the generating cost would be increased from 3.5 to 5.2 p/kWh, compared with 3.5 p/kWh for nuclear.

The general conclusion from this discussion is that the non-environmental costs of coal and nuclear are rather similar, with oil possibly somewhat higher. The differences are attributable to different conditions such as the proximity of the fuel in the case of coal and oil, and to accounting procedures. The environmental costs however differ widely, as will be shown in the next chapter, so the decisions must be made on these grounds, not on that of cost.

5.4 Safety

There is no completely safe way to produce energy. Coal mining is notoriously dangerous, oil wells catch fire, tankers collide or explode and dams burst. The renewable energy sources are sometimes described as safe or benign, but if we take into account the risks involved in making the materials, and then constructing them, it turns out that they are not so safe after all.

To assess safety objectively we must take into account all the risks for each energy source. Mining, transport, construction, operation, maintenance and distribution all involve risks. Some are direct and affect the workers, and some are indirect like pollution and affect the whole population. To compare the sources of energy these risks must be added together. The results of two studies are shown in Figs. 5.2 and 5.3. The former gives the risks associated with different energy sources in the form of deaths involved in the production of a thousand megawatt years of electricity in different ways. Rather similar results are obtained for the injuries to the workers. There is much argument over these estimates, and the results should not be regarded as final. However further studies are likely to give the same ranking order for the various energy sources.

At first sight some of these results might seem surprising. Coal is very hazardous because of the dangers of mining and also because of the large amounts of sulphurous and nitrous gases discharged into the atmosphere.

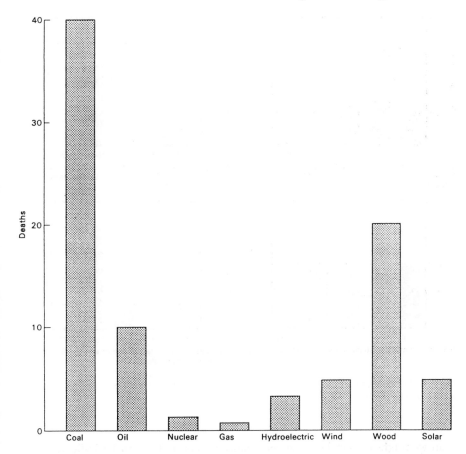

Fig. 5.2. Average numbers of deaths associated with the production of 1000 MWy of electricity in different ways (Inhaber, 1981).

More than 1500 seamen lost their lives in tanker accidents from 1968–79. In December 1989 the Philippines Ferry Dona Paz collided with an oil tanker and more than 1500 people were burned to death. From 1976 to 1982 there have been 1102 deaths due to oil tanker accidents. Gas leaks from pipelines can cause fires and explosions. Poisonous leaks and domestic gas explosions caused 39 deaths in the UK in 1989 alone; this may be compared with five in the nuclear industry (not due to radiation), 79 on off-shore oil and gas installations and 366 in coal mining. Wind and solar are more dangerous

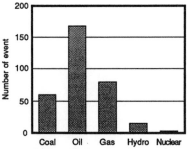

Total number of energy-related severe accidents (1969-1986). Oil and gas dominate. The nuclear figure is one - Chernobyl.

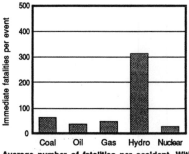

Average number of fatalities per accident. With the large loss of life that can occur in any dam failure hydropower dominates.

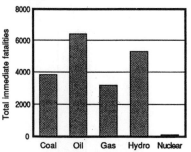

Number of immediate fatalities resulting from accidents.

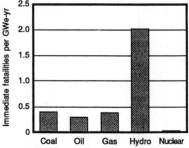

Immediate fatality rate per unit of energy. Here again hydropower dominates.

Fig. 5.3. Accident statistics for various energy sources (Paul Scherrer Institute 1995; *Nuclear Issues*, April, 1982).

than we might expect because very many collectors have to be built, and this means mining and construction hazards. Nuclear comes out quite well because uranium is such a concentrated power source that relatively small quantities need to be mined, and also because there is no poisonous smoke from nuclear power stations.

It is notorious that it is the spectacular accidents that make the headlines and capture the public imagination, although over a long period of time (apart from dam bursts) they contribute relatively little to the total hazard of a power source. This is very familiar in other contexts; a train crash causing some tens of deaths receives more attention than thousands killed individually in road accidents.

Such spectacular accidents are also a tragic feature of energy generation and some of the most notable in the period 1969–1986 are listed in Table 5.5.

Table 5.5. Severe accidents 1969–1986.

	Number	Type	Deaths per accident	Average per year
Coal	62	Mining	10–434	> 200
Oil	63	Fire	5–500	50
Gas	24	Fire	6–452	> 80
Hydro	> 8	Dam burst	11–2500	> 200
Nuclear	1	Chernobyl	31	—

Mining accidents are particularly serious, and over 80,000 miners were killed in accidents from 1873 to 1938, and many hundreds of thousands more had their health permanently impaired by silicosis and other diseases.

The hazards of energy production can of course be reduced by increasing the safety precautions, but this inevitably increases the cost. For example, many of the poisonous gases can be removed from the effluent gases from coal power stations by installing special equipment, but this is very expensive. Furthermore, the safety equipment itself has to be manufactured, and this introduces further hazards. There comes a point where more safety devices have the net effect of reducing the overall safety. In the end we have to strike a balance, and this can only be done by carefully evaluating the hazards of all the energy sources.

We know a great deal about radioactivity because extremely small amounts can be detected by special instruments. In this way we can find out how it travels through the atmosphere, the rivers and the seas, and our own bodies. Very detailed and extensive studies of nuclear hazards have been made, and this allows strict limits to radiation doses to be established and enforced, as discussed in Sec. 3.3.

Much less is known abut the other hazards of life, particularly those associated with the harmful chemicals released into the atmosphere by coal power stations and many factories. This may travel thousands of miles and fall as acid rain. Other chemicals are released into the rivers and the seas. We are all familiar with the black smoke that hangs over industrial cities, and with polluted rivers and lakes. All this exacts a heavy toll in disease and in shortening of life, as well as killing trees, plants and fishes. This is discussed in more detail in Sec. 6.2.

To sum up the comparison between the different energy sources so far, we have seen that only coal, oil and nuclear have the capacity to provide the huge amounts of energy needed by large cities and modern industrial societies. Hydropower is important but limited by the availability of suitable rivers. The renewables like wind and solar have useful small-scale applications, but cannot provide power on the scale we need.

5.5 Reliability

For most industrial and domestic uses, particularly when electricity is being used, it is essential that the supply is reliable. The use of computers is rapidly spreading, and an unexpected interruption of electrical power immediately brings many vital activities to a standstill. Industrial processes controlled by computers stop, electronic communications are cut and supermarkets have to close. If the breakdown is prolonged, households relying on electricity for heating as well as lighting are in severe difficulties in cold weather, and the whole economy suffers. We are so used to relying on our power supplies that we are thrown into disarray when they fail. In New York a few years ago, one power station was overloaded and automatically cut out, thus throwing more load on another, which cut out in turn and so on until the whole electrical grid was down. The grid system was so complicated that it took several days to restore the electrical power. Such events are very rare, but power breakdowns are everyday occurrences in many countries, so much so that it is not possible to carry out many activities, such as lengthy computer calculations or delicate manufacturing processes.

It is neither possible nor necessary for every power station to be able to operate indefinitely, but they should normally operate at least for several months at a time. This is the case for coal, oil, nuclear and hydroelectric power stations. Their electrical output is fed into a national grid, so if one power station breaks down the load can be carried by the other without interrupting the supply. It is also necessary to disconnect power stations from time to time for maintenance, and the grid allows this to be done without difficulties.

The demand for electricity fluctuates, but usually in a predictable way. The greatest demand is in the winter, so maintenance shutdowns are scheduled for the summer. The greatest demand come during a sharp cold spell, so the maximum number of power stations are made available at times when this may happen. It may be necessary to increase power output rapidly, so planning must take account of the time it takes for a power station to come up to full power. Nuclear power stations are very reliable, but take many hours to switch on, so they are operated continually to provide the base load. Gas power stations can be switched on rapidly, so these are kept on standby for sudden increases in demand.

The renewable energy sources vary in their degrees of reliability. Hydropower is generally very reliable, though it can fail in times of prolonged drought. Wind and solar are clearly unreliable, and this is a crippling and unavoidable disadvantage. The electrical grid system can mitigate this unreliability to some extent, since the wind may be blowing in one part of the country but not in another, but not sufficiently to be of practicable use.

There are some applications where reliability is not important. For example, a farmer may need to pump water up to a storage tank, and this can easily be done with a wind pump. Providing the tank is large enough, there is always enough water when needed, and it is topped up whenever the wind is blowing. Such wind pumps are a familiar feature of the landscape in countries with inadequate rainfall. A similar application is the use of solar power to dry hay.

Among the other renewables, tidal power is very reliable, but it only provides energy for part of the time following high tide; this is predictable but has the disadvantage that the tides follow the moon and not the sun. Wave power is reliable but impracticable.

The unreliability of some of the renewables would not be a serious problem if there were a way of storing energy on a large scale. Energy could then be generated when the wind is blowing or the sun shining, and then stored for use whenever required. This is in fact what is done in wind pumps on a small scale, but it is not practicable on a large scale because of the large amounts of water that would be required, not to mention the economic aspects. One possibility that has some chance of success is when there are two substantial lakes at different heights above sea level and not too far from each other. When there is excess supply, water is pumped from the lower lake to the upper and, when power is needed, the water from the upper lake is used to generate electricity by a hydropower station. In spite of the energy losses that occur whenever electrical energy is transformed to potential energy and back again this is more economical that building extra power stations to cope with the peak demand. Unfortunately there are few suitable pairs of lakes for this to be generally useful. There is such an installation at Dinorwig in Wales. This is designed to make the existing grid of conventional power stations more efficient, but it could also be used for the renewables. This would however be practicable only for renewable generators in the vicinity of the lakes due to the high cost of transporting electricity.

Electrical energy can also be stored chemically, for example in batteries, but this is uneconomic on a large scale due to the high cost of the batteries. Batteries are only useful on a small scale where economic considerations are relatively unimportant.

The result of all this is that unreliability of some of the renewables, particularly wind and solar, makes them unacceptable as a major source of energy. They are however useful in small-scale applications where reliability is not important. In other applications their use depends on other factors such as cost, safety and effects on the environment.

5.6 Comparison of Energy Sources

The development of energy sources throughout history is marked by a continual progression from the less concentrated to the more concentrated. By concentration we mean the calorific value per unit of mass. The more concentrated a fuel, the easier it is, in general, to obtain the energy we need, particularly if it has to be transported over large distances. Among the fossil fuels wood is the least concentrated, then comes coal and finally oil is the most concentrated. Nuclear is far more concentrated: the heat obtainable from a pound of uranium is more than can be obtained from a ton of coal.

By contrast, the renewable energies are far less concentrated. The total energy in the earth, the winds, tides and in sunlight is enormous, millions of times of the world's energy needs. But it is so thinly spread that we have to go to a great deal of trouble, and therefore expense, to collect it. We would have to drill many deep holes in the earth to extract the geothermal energy; we have to build many windmills to capture the energy of the wind, and many solar panels to obtain that of the sun. It is only in the case of hydroelectric power when the valleys concentrate the energy for us, that a renewable source is able to provide energy economically on a large scale.

There is another way that energy sources can be classified according to general physical principles. The first category relies on moving material, such as the air in wind power and water in wave power. These sources are the least efficient. The next category contains the sources that obtain their energy by chemical reactions; these sources include wood, biomass, coal and oil, and are more efficient. Finally there are those using nuclear reactions, particularly fission and fusion, and these are the most efficient.

The discussions of the possible energy sources according to their capacity, cost, safety and reliability show that when the oil and natural gas become exhausted there are only two sources able to meet a major part of world needs: coal and nuclear. Hydroelectric can make an important contribution, but it is essentially limited. The other renewable sources can contribute only at the level of a few per cent. This is not due to lack of technical investment but to basic unalterable physical facts. It may be possible to improve their efficiencies in various ways, but there is no way of removing the limitations of low power density and unreliability.

6

Effects on the Environment

6.1 Energy and Health

Our health and well-being and our standard of living depends very directly on our environment. By our environment we mean our surroundings, the air we breathe, its purity and its temperature, and the noises and other disturbances to which we are subjected. We mean the food we eat, the comfort of our homes and the availability of convenient transport and communication. We include also the appearance of the city or town where we live, and the beauty of the surrounding countryside. All these determine our bodily and mental health.

Much of this environment depends on energy, and so energy, by contributing to our environment, also affects our health. The relation between energy and health is shown by Fig. 6.1. The vertical scale is the average life expectancy, a crude but effective measure of health, and the horizontal scale shows the average annual consumption of energy by each person. Each dot shows the relation between life expectancy and energy consumption for a particular country. In the lower left-hand corner are the poorer countries of the world, with little energy and low life expectancy. At the top right hand are the rich countries of Europe and North America and Japan where each person uses up tens or hundreds of times the energy per person of the poorer countries, and where the life expectancy is about twice as great.

This shows that if we want to increase the standard of living of the poorer countries of the world we must increase the amount of energy available to them. Unfortunately, however, the energy sources themselves, the power stations and the associated industries, all have a deleterious effect on the environment. It is therefore important to examine the effects on the environment of the various

HEALTH

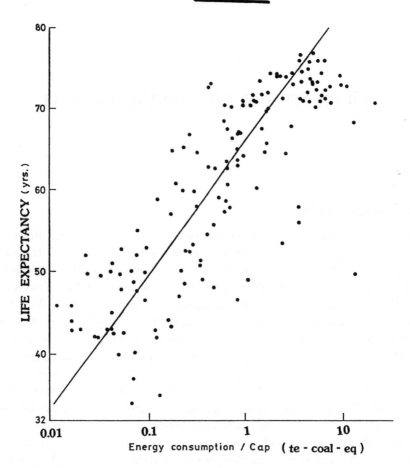

Fig. 6.1. Life Expectancy 1980–1985 as a function of energy consumption per head in 150 countries. Each country is represented by a point.

energy sources in order to decide which of them are preferable from this point of view, and how their environmental impact can be minimised.

Although every energy source, and indeed every industrial process, produces unwanted and often dangerous waste products, these can sometimes be processed so as to become quite safe. We therefore define pollution as the production of waste products that cannot be re-integrated into the earth's ecosystem by natural or industrial processes at a rate comparable with that of their production.

Some of these effects are damaging to our health and others are destructive of the landscape. The more energy we produce, the greater the environmental effects. Thus, paradoxically, as we generate more energy to improve our living standard, we cause pollution that reduces our living standard. Sooner or later, unless we are very careful, we will do more harm than good.

In this chapter, the pollution of the atmosphere is described in Sec. 6.2 and that of the earth in Sec. 6.3. The visual effects on the landscape and aural pollution are considered in Sec. 6.4 and electromagnetic pollution in Sec. 6.5. The effects of global warming and the greenhouse effect are considered in Sec. 6.6, and chemical pollution in Sec. 6.7.

6.2 Pollution of the Atmosphere

A major effect of energy generation on the environment is the pollution of the atmosphere and the earth. All power stations pollute in one way or another, and the more power we generate, the greater the pollution. In some cases the pollution is an inevitable result of the process itself and therefore cannot be prevented, while in others it can be controlled so that the harm to people is minimised

Pollution of the atmosphere first occurred on a large scale in the cities of the industrial revolution in the nineteenth century. We are familiar with pictures of factories belching black smoke over streets of grimy houses. The London of Sherlock Holmes was often shrouded in fog from millions of domestic fires. As recently as 1952 atmospheric inversion trapped the smoke and produced a dense London fog that killed over three thousand people. The availability of gas and electric heating has made it possible to prohibit domestic fires, eliminating this form of pollution. In other countries, however, wood and dung are still used on a large scale for domestic heating and cooking, and cities like Calcutta, for example, are seriously polluted.

Coal power stations are particularly polluting, and a typical one will emit each year eleven million tons of carbon dioxide, sixteen thousand tons of sulphur dioxide, twenty-nine thousand tons of nitrous oxide, a thousand tons of dust and smaller amounts of a whole range of other chemicals such as aluminium, calcium, potassium, titanium and arsenic. To produce 1 GWyr of electricity about 3.5 million tons of coal is burnt, and this coal contains about 5.25 tons of uranium. Most of this is caught by the filters, but a few thousand tons of ash will escape carrying with it a corresponding fraction of the uranium. This accounts for the radioactivity emitted by coal power stations. All of this is poured forth into the air we breathe, and inevitably it damages our health. It travels large distances and falls as acid rain.

The main components of acid rain are sulphur dioxide, the nitrogen oxides and hydrocarbons. Sulphur dioxide damages limestone and marble and some types of sandstone, and so is harmful to buildings. Together with nitrous oxide it can be lethal to fish. Already this is affecting lakes, rivers and forests in many countries, including Scandinavia, Austria, France, Switzerland, Britain and the United States. The forests of central Europe, particularly in Germany, are badly affected; a survey showed that 560,000 hectares, about 8% of German forests, are seriously deteriorating. The acid in the water can dissolve copper and lead from pipes, and cadmium and mercury from the soil, and all this is hazardous to health. More sulphate particles in the air are increasing the incidence of asthma, allergies and hay fever. It is possible in principle to remove some of these poisons before they are expelled into the atmosphere, but to remove them all would be prohibitively expensive.

These noxious gases are emitted in many industrial processes, but power plants contribute a large share. Analysis of emissions in the United Kingdom showed that 66% of the sulphur dioxide comes from fossil fuel power stations. About 46% of the nitrogen oxides come from the same source.

It has been estimated that the annual death rate in the United States from air pollution is about 60,000 per year, and 10,000 per year in the United Kingdom, largely due to road transport and the burning of coal. About 75 millions of tons of sulphur are discharged into the atmosphere each year, mostly due to the burning of fossil fuels, and this is a major cause of death due to the small particles that can lodge deep in the lungs. It is possible to eliminate most of the sulphur emissions by fitting desulphurisation equipment, which is necessary because British coal has an average sulphur content of 1.6%. This equipment costs over £100M for each power station, and adds about £30M to the operating costs, not including those of mining 300,000 tons of limestone and removing 16,000 tons of calcium sulphate and 500,000 tons of gypsum for a 2000 MW power station. Already about £5B has been spent in the UK on this equipment, and the cost could rise to £8B. A study by the International Institute for Applied Systems Analysis estimates that the annual cost to the UK of reducing emissions by 80% would be about £1B. This is proving so expensive that coal with a lower sulphur content (0.8 to 1%) is now being imported, and more of the cheaper, easily-built gas power stations are coming into operation.

A further hazard is due to the polycyclic hydrocarbons that come from the incomplete burning of fossil fuels. About two tonnes are emitted from a 1000 MW coal power station each year, and about a thousand cancer deaths per year in the UK are attributed to this cause. In 1986 there were an extra 13,000 cancer deaths per 100,000 in urban areas in England compared with rural areas. Thus has been attributed to air pollution by vehicle exhausts, possibly by the benzo-pyrene in city air.

The renewable energy sources are widely believed to produce no atmospheric pollution. This is true for their everyday operation, but not for their construction. Like all power stations, they have to be made in factories, and this inevitably produces pollution. This is by no means small, as very large numbers of windmills and solar panels have to be made to equal the output of a conventional power station.

Since nuclear power is the only large-scale source, together with coal, that can take over from oil and natural gas, its effects on the environment must be carefully examined. In normal operation, a nuclear power station releases only very small amounts of radioactivity into the atmosphere. These amounts are even less than the radioactive emissions from coal power stations.

To put into perspective the radiations from the radioactive material produced in nuclear reactors they must be compared with that from other sources of radiation, and this was done in Chap. 3. The radiation from nuclear power reactors is kept carefully under control and is currently about a quarter of a millirem per year. The environmental effects of a nuclear reactor accident may be very severe, and these are discussed in Chap. 4.

There is one additional source of atmospheric pollution that deserves mention, although it is only indirectly connected with energy production. The uranium present in the earth decays gradually, producing the radioactive gas radon. Radon is responsible for almost half of the natural background radiation, as shown in Table 3.1. This radon seeps upward through the ground, and is soon dispersed by the wind. If however it is trapped, as it may be inside a house that is not well-ventilated, the radon may accumulate until the radiation level is comparable with the Governmental recommended limit of 200 Becquerels per cubic metre, corresponding to about 10 mSv/yr. It is notable that this limit is ten times the limit of 1 mSv/yr imposed on doses to members of the public from nuclear power stations. A survey in 1985 by the National

Radiological Protection Board showed that the radon concentrations in London are half the national average, while in south-west England they are about three times average. Studies in Canada, China, Finland, France, Germany, Japan, Sweden and the USA showed no correlation between domestic radon exposure and lung cancer.

Some houses in regions like Devon and Cornwall where there are granitic rocks have radon levels fifty times the national average, corresponding to an annual dose of about 20 mS, that is ten per cent of the average natural radiation background level over the whole country. However, lung cancer rates in these two countries are well below the average national level. Furthermore, there is no correlation between leukaemia rates and radon levels in various parts of Devon and Cornwall. It is very easy to remove the radon by increased ventilation, but this is contrary to the policy that requires us to conserve energy by increasing insulation and shutting all doors and windows. So, if we live in an area with high radon levels, we may have to decide between energy conservation and radon irradiation. It should be added that this is not a serious problem, since the radon levels are far below those likely to cause any appreciable harm.

6.3 Pollution of the Earth

All methods of energy generation, and indeed all industrial processes, inevitably produce waste. This waste is nearly always dangerous, and sometimes extremely dangerous. With increasing industrialisation this waste has become a serious world-wide problem.

The character of the waste, and the means adopted to dispose of it safely, is very different for the various energy sources. Wood is generally burned on a small scale, and produces some ash, which is easily dealt with.

Coal power stations produce huge amounts of ash due to the impurities in the coal. Most of this is expelled into the atmosphere as smoke, and poisons the air we breathe. Each coal power station produces each year over a million tons of ash, together with five hundred thousand tons of gypsum and twenty-one thousand tons of sludge. Although this material is not particularly hazardous, its disposal is a substantial operation due to the huge amount of waste. The ash is frequently dumped in the sea where it destroys the habitat of crabs and lobsters, and poisons fish. It is interesting to note that the alpha-particle radioactivity of the ash is often 1000 Bq/kg for a power station in Northumberland. If this had come from a nuclear plant it would be classified as low-level waste, for which the limit is 400 Bq/kg, and so would have to be carefully buried at a cost that would render coal power stations uneconomic.

Oil power stations are not quite so polluting as coal power stations, but they inevitably produce the same amount of carbon dioxide. Oil does have an additional hazard due to the need to transport it over large distances. Overland this can be done by pipelines, and these affect the environment not only visually but due to the possibility of leaks. The seasonal migration of animals such as caribou may also be affected. Much oil is also transported across the oceans by huge supertankers. This is very efficient, but if they run onto the rocks and are holed or break up large quantities of oil are released. This oil floats on the water and can travel large distances, causing a major environmental disaster. The Torrey Canyon, Exxon Valduz and quite recently the Sea Empress are still vivid memories. The oil fouls the beaches and kills birds, marine life and fishes. The oil can be dispersed by spraying with chemicals, but in some cases this does even more environmental damage. Although some of changes may be irreversible, the environment often recovers remarkably quickly, so that some years later few traces of the disaster remain. As in other matters, it is important not to overstate the extent of the damage, as is often done by environmental organisations and by those who justifiably seek compensation. Oil tankers are subject to stringent regulations, and improvements in their design are continually being made. Nevertheless these disasters still happen, and remain a serious environmental hazard of oil power.

Carbon dioxide is inevitably produced by coal and oil power stations since it is the combination of carbon and oxygen that generates the heat energy. This carbon dioxide is responsible for the greenhouse effect, which is considered in Sec. 6.6.

The radioactive waste from nuclear reactors has been discussed in Sec. 3.4, together with the methods used to ensure that it is buried deep underground so that it cannot cause any harm. If these procedures are carried out, there is no harmful pollution of the earth. Unfortunately, however, the necessary care is not always taken. Thus one of the worst examples of radioactive pollution is to be found near Murmansk on the Kola peninsula in northern Russia. After the end of the Cold War, over a hundred nuclear submarines, containing two nuclear reactors each, were dismantled and tens of thousands of used fuel rods stored in a rusting ship or left lying on the quayside. Concrete has been poured over the fuel rods in the ship, making it top heavy and liable to capsize in a storm. All this material is highly radioactive and extremely dangerous. The radioactive fuels rods can be treated and stored safely, but how can we be sure that this is always done?

The concern about waste from nuclear reactors is not matched by a similar concern for industrial waste, which is often radioactive. Thus the spoil from old mines in Devon and Cornwall was often dumped in any available place and can contain appreciable amounts of uranium. Such uranium-bearing spoil is scattered over popular beaches and recently a rock containing 1 kg of uranium was found just below the surface. If that rock had come from the nuclear industry it would have been classed as intermediate level waste and buried deep underground. Uranium often occurs together with blue copper minerals or red jasper, so they are likely to be collected for their appearance. It is easy to find rocks with specific activities of 10,000 Bq/gm. According to IAEA regulations, anything with a specific activity exceeding 74 Bq/gm should be transported in a vehicle with warning placards. Even the roads are radioactive; levels up to 9 μSv/hr have been recorded. In some mine workings, open to the public, the radiation dose is 0.7 mSv/hr, which may be compared with the ICRP limit for radiation workers of 20 mSv/yr. Sometimes the air in the mines has radon concentrations of 7 to 10 million Bq per cubic metre; NRPB recommends action to remove radon from homes if the level is above 200 Bq/cm.

Workers in many industries are exposed to radiation. The highest annual dose of 14 mSv is received by miners not in the coal industry. Second comes the nuclear industry and air crews, with 2 mSv, then coal miners with 1.2 mSv. The NRPB recommends maximum dose of 15 mSv averaged over four years. The large number of people who work indoors with high radon levels incur a collective dose of 3000 man-Sv, far exceeding all other industrial exposures put together. Radioactive waste contributes only 0.04% to the 100 million tons of toxic waste produced annually in the UK, and almost all of this is low level radioactive waste.

Hydropower produces no waste, but severely affects the environment by inundating huge areas. The Parana dam in South America, the Volta dam in Ghana and the Aswan dam in Egypt, to mention just three examples, have inundated tens of thousands of sq.km. and this inevitably has great effects on the environment. In Europe, rather smaller dams frequently inundate picturesque mountain valleys, and displace people from areas that have been cultivated for thousands of years.

Biomass also pollutes the earth. For each litre of alcohol produced from sugar cane there are 12 to 14 litres of polluting effluent. It also uses up agricultural land that could be used for food production.

The two major sources of energy for the future, coal and nuclear, thus produce very different types of waste. That from coal power stations goes into the atmosphere as well being produced in solid form. Nuclear power stations, however, discharge almost no waste into the atmosphere; all the waste is in the radioactive fission fragments. The solid waste from both types of power stations can be safely dealt with, but the atmospheric waste due to coal and other fossil fuel power stations is a most intractable problem with most serious consequences for health and for the environment.

6.4 Visual and Aural Pollution

The landscape usually means nature in its original state, untouched by man. This is often of great beauty, and we want to do all we can to preserve it. Over most of the earth, however, man has already left his mark, and it is not always destructive. In many countries the work of farmers and builders over the years has created a man-made landscape that also has great beauty. An example is the Cotswold country in England, with its gently rolling hills, its well-tended farms, its flocks of sheep and its golden-stone villages and magnificent parish churches.

We want to prevent this from being spoiled by ugly buildings, factories, electric pylons and motorways, so first of all we consider the visual impact of energy generation.

All power stations are large buildings that have a great visual impact on the landscape. Coal power stations are usually surrounded by huge piles of coal waiting to be burnt and by heaps of discarded ash. Oil and nuclear power stations are somewhat smaller, but all have huge cooling towers to reduce the temperature of steam after passing through the turbines and thus to increase their efficiency. It is possible, to some extent, to site such power stations to reduce their visual impact, but in general they are blots on the landscape. The main consolation is that each of them produces such a large amount of power that a rather small number suffices for the needs of a whole country.

Electricity pylons are now a familiar feature of the landscape, and are inevitably associated with power generation. It is possible but unfortunately prohibitively expensive to bury the power lines underground. This may however be overcome by new technologies such as high temperature superconductors.

Hydroelectric power is extremely destructive of the landscape. To ensure a continual supply of water, the rivers are dammed to form large lakes, and this usually means that beautiful mountain valleys are destroyed. This is a major tragedy for the people who have lived there for generations, as has happened in many cases, for example in the European Alps. The lake formed by the dam may have some amenity value, but during the year the water level rises and falls, and in the dry season this exposes ugly sterile bands of mud. There is now increased public awareness of the destruction of the environment due to hydropower, and environmentalists now make strong protests. A recent example is provided by the dams proposed by hydroelectric companies in Western Tasmania, which threatened to destroy an area of outstanding natural beauty and unique scientific interest. Energetic protests were only partly successful, but some of the area has been saved.

Hydropower is more dangerous than is usually believed. Dams may look solid, but they can be fractured by earthquakes and undermined by water seepage. If they burst, they cannot only cause huge loss of life, but further damage to the environment.

The renewable energy sources have even greater effects on the environment. The energy they tap is very thinly spread, and so inevitably the collectors must occupy a very large area. So far only wind power has been developed on a moderate scale, in the form of wind farms. To catch the available wind, they must be placed on high ground, where they can be seen for many miles. Many hundreds of large windmills are needed to give the same output as a coal or oil power station, so the wind farms occupy very large areas. Each windmill requires an access road, and this interferes with agriculture and increases the cost of returning the land to its previous state when the windmill is dismantled.

Another type of pollution is aural, and this is characteristic of wind power. Most power stations emit some noise, but this does not matter if the power station is on a relatively restricted site. Windmills, however, are spread over large areas and emit a persistent humming noise that is very disturbing to anyone living nearby. People who go to live in remote areas usually do so for peace and quiet, and their lives are soon ruined. Naturally they are reassured that the windmills are silent, but anything that rotates eventually wears down the bearings and gets progressively more noisy. They may decide that they must move to a quieter area, but are not able to do so because they cannot sell their house because no one else wants to live there.

The other renewable sources, if they are developed, are also likely to have serious effects on the environment. Tidal power require a huge barrage across an estuary such as that of the river Severn, and this changes the whole ecology of the area. If this were built, the lives and perhaps the very existence of the fishes, the birds and the plants of the area would be irrevocably affected.

Wave power requires large structures along the coastline, and these would be visually prominent and disruptive to sailing, swimming and other beach activities.

Some recent figures for the amounts of land in square metres taken up by various forms of power stations for each megawatt of energy generated are: nuclear 630, oil 870, gas 1500, coal 2400, solar 1,000,000, hydroelectric 265,000 and wind 1,700,000. It should be added that in the case of wind power the turbines are so far apart that much of the land between them can still be used for agriculture.

6.5 Electromagnetic Pollution

One of the lesser-known effects of living in a technological society is that we are all exposed to electric and magnetic fields in addition to those occurring naturally. There has been much discussion about whether living near electric power lines affects health, and in addition there are the possible effects of the electromagnetic fields generated by home appliances such as electric bells, lights, television sets, radios, microwave ovens, electric shavers, drills and so on. Some studies have shown an increase in the number of cases of childhood leukaemia among people living near power lines, but other studies have shown no significant effect. There have been many legal actions but without conclusive evidence none has been successful.

The situation is similar to the question of leukaemia clusters around nuclear plants, and can be tackled in similar ways. In the first place, a correlation, even if established, does not imply a causal influence; there could be some other cause of both of them. This can be studied by looking at the incidence of leukaemia clusters in other circumstances. Secondly, the intensity of the radiation can be compared with both the natural background and the artificial background due to household appliances. If the electromagnetic fields due to power lines are much weaker than those attributable to home appliances then it is very unlikely that the power lines are to blame. Thirdly, it is necessary to understand the mechanism by which the electromagnetic fields induce leukaemia.

The intensity of the magnetic field due to a 400 kV overhead power line is about 3 μT (microtesla) out to a distance of 10 m and thereafter falls rapidly to 100 nT at 100 m. The UK Radiation Protection Board has established an exposure limit of 1.6 mT based on the fields that can induce electric currents similar to those produced naturally by nerve and muscle action; those due to power lines are thus about two thousand times less than this limit. From 100 m to 200 m the field due to a power line is similar to that in a typical house due to various electrical appliances, namely from 10 to 100 nT. If we do not worry about these fields, then there is no need to worry about the fields of power lines.

Many possible mechanisms have been proposed for the effects of electromagnetic fields on our bodies, but none of them are acceptable.

Thus at present there is no good evidence for believing that power lines are injurious to health. Research should continue, but until definite evidence is forthcoming it is not a cause for concern.

A rather similar concern is the effect of the electromagnetic fields associated with mobile telephones. It has recently been reported that they can cause cancer, but the evidence is inconclusive. This is another example of an alleged connection that would be serious if it were definitely established. Research must continue, but it is a delicate political problem to decide how much to spend on such work and how much on alleviating the consequences of other cases such as the association between smoking and lung cancer where the reality of the effect has been established beyond doubt.

6.6   Global Warming and the Greenhouse Effect

Another much more radical way that the landscape could be destroyed is by climate change. If the average world temperature varied so that the balance of the ecology was upset and plants and trees died, or if the land itself were inundated, that would indeed be the destruction of the landscape. It is already certain that increased energy production is gradually heating the atmosphere and that plants and trees are being affected by acid rain, but far more drastic changes could be caused by the greenhouse effect.

Even if we were able to burn pure coal, so that there would be no ash or poisons in the smoke, we would still inevitably produce carbon dioxide. This is the result of burning the coal, so that the carbon in the coal combines with the oxygen in the air to give carbon dioxide. The anthropogenic production of carbon dioxide amounts to about eight gigatons of carbon per year, with

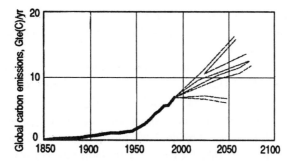

Fig. 6.2. Rapid growth in global carbon emissions and possible future scenarios. A contribution from nuclear energy is needed to stem the growth (*Nuclear Issues*, September 1996).

contributions of 33% from oil, 28% from coal, 15% from gas, 20% from land-use change and smaller amounts from other sources. This can also be classified as 37% from power stations, 20% from industry, 16% from road transport, 14% domestic and 13% other sources. Over the last few decades, it has been found that the amount of carbon dioxide in the atmosphere is steadily increasing as shown in Fig. 6.2. Much of it is removed by plants and trees as they grow, and more is absorbed by the oceans. However the destruction of forests in so many countries and the large amounts of carbon dioxide produced by power stations and other industrial processes has led to a net increase. It is estimated that by the middle of the next century that amount of carbon dioxide in the atmosphere will be twice what it was in pre-industrial times.

The effect of this increase in the concentration of carbon dioxide in the atmosphere is to retain more of the sun's heat, just like a greenhouse. It might at first be thought that this would be a good thing, we would all be warmer and so need to produce less energy. However it is not as simple as that. If the earth warms up, part of the Antarctic ice cap will melt, raising the sea level, and so flooding low-lying countries like Bangladesh and Holland. The melting of the Arctic ice does not raise the sea level because it is already floating. It is very likely that the rise in temperature will also seriously affect the world's weather, with incalculable results. Rainfall patterns may change and storms disrupt commerce and cause local devastation.

In addition to carbon dioxide, there are several other gases that contribute to the greenhouse effect, in particular methane, nitrous oxide and the chlorofluorocarbons (CFS) that are used in refrigerators. The last two of these are far more damaging per molecule than carbon dioxide, by factors of about 200 and 4000, but the quantity of carbon dioxide is so much greater that it accounts for more than 60% of the greenhouse effect. The very great damage caused by the CFS had led to demands that they be banned as soon as possible. It is desirable, however, that this should not be done until an economic substitute is found, as refrigerators are very useful, especially in hot countries.

The concentrations of these anthropogenic greenhouse gases are increasing annually by 0.4% for carbon dioxide, 1.2% for methane, 0.3% for nitrous oxide, 6% for CFS and about 0.25% for ozone. The main uncertainty is the effect of water vapour, which is greater than that of all the other gases combined. It is determined by the climate system leading to changes in the cloud cover, which in turn changes the amount of solar energy absorbed or reflected. The climate predictions are thus very sensitive to the amount of cloud cover.

There is much discussion about the time-scale and the magnitude of the global warming and of its effects. The whole question has recently been studied by the Intergovernmental Panel on Climate Change, involving several hundred scientists from a hundred countries meeting under the auspices of the World Meteorological Organisation and the United Nations Environment Programme. While there are many uncertainties, the following figures give the best available estimates.

It is established that the mean surface air temperature has increased by from 3 to 6 degrees C over the last hundred years, and the five warmest years have all occurred in the 1980s. It is estimated that by the year 2100 the temperature will have increased by another 4°C if nothing is done to reduce emissions, and by about 2°C if emissions are controlled. If nothing is done, the rise in temperature is estimated to increase the sea level by 60 cm by 2100, or by about 40 cm if emissions are controlled.

It is possible that the recent floods in Europe and the USA and the drought in Africa are early manifestations of the greenhouse effect. The very warm summers in recent years greatly reduced grain yields. It is however very difficult to be sure that this is due to the greenhouse effect, because the weather fluctuates widely from one year to another. The earth's ecosystem is to some extent self-correcting, so it is also possible that the effects are being masked by other changes. It is likely to be some years before we are absolutely sure that the greenhouse effect is real, but if we wait until then, it will be too late to do anything about it. There is thus strong international pressure on all countries to reduce their carbon emissions by pledging a percentage reduction each year.

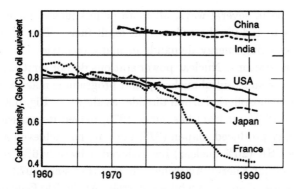

Fig. 6.3. Carbon intensity: the rate of carbon emission per unit of primary energy use in various countries (*Nuclear Issues*, September 1996).

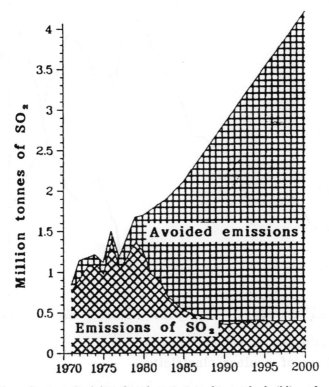

Fig. 6.4. The reduction of sulphur dioxide emissions due to the building of nuclear power stations in France (*Nuclear Issues*, April 1982).

This can be done in several ways, all more or less inconvenient. By far the simplest way is to reduce the reliance on fossil fuels. This is however very difficult because rising energy demands in many countries are being met by coal or oil power stations. Nuclear power stations do not emit carbon dioxide, so countries with a substantial nuclear power programme make a much smaller contribution to atmospheric carbon dioxide, as shown in Fig. 6.3.

Thus France (80% nuclear) has, since 1970, halved its emission of carbon per unit of energy produced. Japan (32% nuclear) has achieved a reduction of 20%, while the USA (20% nuclear) has reduced it by only 6%. A study made of the US Council for Energy Awareness showed that since 1973 nuclear power has reduced the use of oil by 15.5 billion barrels. At the Rio Conference Britain

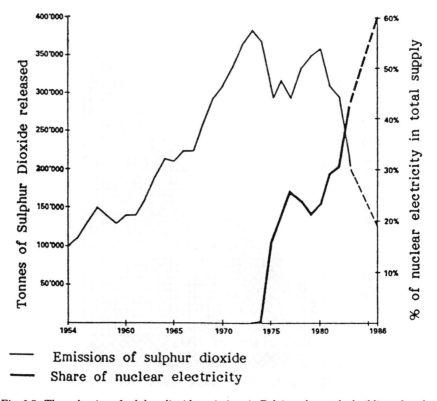

———   Emissions of sulphur dioxide
———   Share of nuclear electricity

Fig. 6.5. The reduction of sulphur dioxide emissions in Belgium due to the building of nuclear power stations (*Nuclear Issues*, September 1983).

promised to reduce carbon dioxide emissions by 10 million tonnes per year in order to reduce the total emission to the 1990 level by the year 2000. This can be done by building two and a half nuclear power stations like Sizewell B to replace existing coal power stations. It is entirely practicable to reduce carbon dioxide emissions in this way, unlike the alternative suggestions that have been made. The emission of sulphur dioxide is also strongly reduced, as shown in Figs. 6.4 and 6.5. In spite of this, three new coal power stations are planned, but no nuclear power stations.

The savings of carbon dioxide emission is illustrated by some figures from 1988. In that year, 1866 billion kWh of electricity were generated by nuclear power stations. The same amount would be produced by burning 900 million tons of coal or 600 million tons of oil. Thus the emission of 3000 million tons of carbon dioxide is saved by using nuclear instead of coal or oil.

The British Government has now set a target of a 10% cut in carbon dioxide emissions in the period from 1990 to 2010. By 1995, as shown in Fig. 6.6 a cut of about 6% had already been achieved. This reduction, however, is due to the increase in nuclear output by 39% from 1990 to 1994. In subsequent years, if no more nuclear power stations are built, the rate of carbon dioxide emissions is set to rise quite steeply. The rate of rise depends on the numbers of coal and gas power stations built; these correspond to the various scenarios shown in the Figure. The increase will be 70 mt from coal and 30 mt from gas. Each nuclear power station will reduce the emissions by 18 mt.

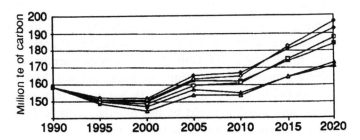

Fig. 6.6. Carbon dioxide emissions in the UK for various scenarios. A million tons of carbon is equivalent to 3.66 million tons of carbon dioxide (*Nuclear Issues*, March 1995).

Building more gas power stations halves the emission of carbon dioxide, but this is offset by the probable leakage of methane, which has a global warming potential 62 times that of carbon dioxide. Estimates indicate that these two

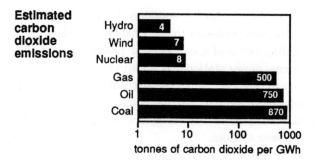

Fig. 6.7. Carbon dioxide emissions from various energy sources (*Nuclear Issues*, January 1995).

effects are of the same order of magnitude and if this is so there is no reduction in global warming to be expected from the switch to gas power stations. The carbon dioxide emissions from various energy sources are shown in Fig. 6.7.

While the more developed countries strive to reduce their carbon emissions, those from the large developing countries like India and China are set to rise from 10% of the world total now to about 50% in 2040.

Carbon dioxide emission can be reduced by imposing a carbon tax. Thus since May 1995 Sweden has imposed a pollution tax of 5 SwK per kg for carbon dioxide, 30 for sulphur and 40 for nitrous oxide. If this were applied in the UK it would add 2 p/kWhr to the price of electricity from coal. A tax on sulphur emission has been proposed in Finland, and this would increase coal prices by 20%, oil by 10% and petrol by 1.5%.

Global warming is not the only possible climate change. It has been suggested that there may be a sudden and catastrophic cooling of northern Europe due to a change in the flow of the Gulf Stream. Global warming could alter this flow by injecting more fresh water into the North Sea and this could cause the temperature to fall by 6 to 8°C. The North Sea would then be frozen for much of the year, and London would be like Siberia. This is an additional argument for reducing the emission of carbon dioxide.

6.7 Chemical Pollution

The pollution from power stations, described in the previous sections, has been carefully studied, but much less is known about many other hazards to life, particularly those associated with the harmful chemicals released into the atmosphere by many factories. These may travel thousands of miles and fall as acid rain. Other chemicals are released into rivers and seas. We are all familiar with the black smoke that hangs over industrial cities and with polluted rivers and lakes. All this exacts a heavy toll in disease and in shortening of life, as well as killing trees, plants and fishes.

By far the greatest danger to the environment from industrial wastes comes from a wide range of manufacturing processes, particularly in the chemical industry. These include poisonous gases emitted into the atmosphere, liquid waste discharged into rivers and seas, and solid waste that is stored on land, often without any secure containment. Corroding drums of cyanide have been found on derelict sites where children play. It has been estimated that the amount of hazardous industrial waste in the United Kingdom alone amounts to between five and ten million tons per year. In Germany, only about 1.5% of the hazardous waste is radioactive. A report by Nriagu and Pacgna lists the annual emissions from industry as 2 M tons/yr of zinc and copper, over 1 M tons of lead, half a M tons of nickel, 100,000 tons of arsenic, molybdenum, vanadium, antimony and selenium, and tens of thousands of tons of cadmium and mercury. This far exceeds the toxicity of all radioactive and organic wastes.

Since the beginning of the industrial revolution, such wastes have been poured into the environment, with very little attempt to reduce the dangers. The managers of factories were under pressure to maximise profits, and they had little thought for the environment. Often the serious effects on the health of their own workers were not understood. Asbestos was widely used as an insulating material before its toxic properties were known, and many people suffered severely. Reducing the volume of hazardous waste and ensuring that it is contained is often very expensive, requiring the installation of new equipment. Even if a company decides to clean up its operations it may face opposition from its shareholders, as the cost would have to come from the profits.

The industrial pollution in Eastern Europe and the former Soviet Union is particularly bad. It might be thought that safety and conservation measures could be more efficiently and fairly enforced in a centralised authoritarian society, but the example of the former Soviet Union shows that this is not necessarily the case. The fifteen republics of the former USSR all adopted laws regulating the preservation of nature, but they are difficult to enforce. The greatest sources of pollution are the large industrial enterprises and these are planned and approved by the central Government, which is not bound by the laws of the individual republics. Decisions concerning industrial development are thus dominated more by the speed of construction and the cost than by considerations of safety and pollution. The local laws are powerless to prevent this and the vocal and influential protest groups that spring up in such situations in a democratic society cannot exist in an authoritarian one.

The results of this are fully documented in a book on *The Destruction of Nature in the Soviet Union* by Boris Kormarov, the pseudonym of an official in a Soviet Ministry. He cites many examples of large industrial developments such as the Estonian shale industry, the Maardu chemical combine, the South Bud atomic power plant, and the Chigirin thermoelectric plan. In each case the local pollution laws have been flouted, and as a result large areas have been rendered sterile and worthless. The damage includes agricultural land covered by slag and sludge from mines and factories, barren forest and swamps left after logging, eroded hillsides, polluted rivers and atmosphere. The total effect is that nearly ten per cent of the habitable land in the former Soviet Union have already been laid waste by pollution, and the process continues at an accelerating pace. This widespread pollution is likely to be one of the causes contributing to the rise in infant mortality and respiratory diseases in the former Soviet Union in recent years (Davis and Feshbach, 1980).

In Poland, the Skawina aluminium smeltery and the Lenin steel mills were built about forty years ago near Cracow for political reasons. Meterorologically, the site is poor because there is little or no air circulation, and as a result the toxic chemicals from the industrial complex have adversely affected the health of the population and the produce of the farms in the surrounding areas.

The effects of such pollution on the lives of the people is illustrated by a recent report from the industrial town of Dzerzhinsk, east of Moscow. There are dozens of factories producing chlorine pesticides and, until recently, chemical weapons. The Kaprolaktam plant was found to emit 600 tons of vinyl chloride, a carcinogenic gas, every year. The children used to play hide and seek in the orange fog, and the sky was always multi-coloured — green, yellow and grey. Due to the economic decline, production has now dropped by two-thirds, and with it the pollution, but the legacy of past pollution is still there. A pond near the Orgsteklo factory gives off a nauseating stench and debris and waste are strewn everywhere. Most of the plants use chlorine and give off dioxins that can enter the food chain and produce cancer, liver disease, skin disorders and damage to the immune and reproductive systems. The life expectancies are 42 for men and 47 for women. Two villages nearby are so polluted that they have been declared uninhabitable and struck off the records so the people who still live there are turned away from clinics and hospitals because they have no address. Russia's leading dioxin expert, Prof. Lev Fyodorov, says that no one should live in the Dzerzhinsk area but the mayor, Alexander Romanov, says that people are only evacuated in emergencies and in any case there is nowhere for them to go. Similar stories could be told of pollution in other parts of the country.

In recent years there has been growing realisation of the health hazards of industrial waste, and strict legal controls are now operating in many countries. Even so, they are not always obeyed, and there remains a strong incentive to reduce the expenditure on safety measures to the bare minimum. In many industrial areas, where factories have been operating for years, the ground is often heavily contaminated with dangerous chemicals produced by factories closed long ago. There have been several cases where houses have been built in such areas, and the health of the people living has been badly affected before the cause was recognised. The cost of cleaning such land is high.

In the last few years there has developed a clandestine trade in unwanted chemical waste between some industrialised and some developing countries. Since the disposal of toxic industrial chemical wastes is now governed by strict regulations, it is often very costly to get rid of unwanted waste. Some developing countries, eager to make some money but not too worried about the hazards of toxic waste, are willing to accept such waste and then just pile it up somewhere. Some of these waste dumps have been found, with stacks of corroding drums leaking toxic chemicals. Needless to say, this is totally illegal and very dangerous, and strong efforts are being made to put an end to such practices.

Factories producing chemicals are also responsible for some of the most serious disasters. That at Bhopal in December 1984 led to the emission of methyl isocyanate that according to the official 1990 estimate killed 3,800 people and injured 203,500. This disaster is now almost forgotten, whereas that at Chernobyl, which killed about 40 people and caused about a thousand cancers, is still remembered.

References

Arvill, R., *Man and Environment*, Pelican Books, 1967.

Blunden, J. and Reddish, A., *Energy, Resources and Environment*, Hodder and Stoughton, London, 1991.

Bocker, E. and Van Grondelle, R., *Environmental Physics*, Wiley, New York, 1995.

Brooks, P., *The House of Life: Rachel Carson at Work*, Houghton Miffin Co., Boston, 1989.

Carson, R., *The Silent Spring*, Penguin Books, London, 1962.

Dasmann, R. F., *Planet in Peril?* Penguin Books, London, 1972.

Davis, C. and Feshbach, M., *Rising Infant Mortality in the USSR in the 1970s*, US Bureau of the Census P-95, No. 4, 1980.

Glacken, C., *Traces on the Rhodian Shore*, University of California Press, 1967.

Houghton, J., *Global Warming: the Complete Briefing*, Lion, Oxford, 1994.

Houghton, J. T., Jenkins, G. J. and Ephraums, J. J. (Eds), *Climate Change*, published for the Intergovernmental Panel on Climate Change, Cambridge University Press, 1990.

Jaki, S. L., *Ecology or Ecologism*, Study Week on Man and his Environment, G. B. Marini-Bettolo (Ed), *Pontificiae Academiae Scientiarum Scripta Varia*, No. 84, 1990.

Komarov, B., *The Destruction of Nature in the Soviet Union*, Pluto Press, 1980, extracts published in the *New Scientist* 88.514.1981.

Nriagu, J. G., and Pacgna, J. M., *Nature* 333.134.1988.

Prance, G., The Earth under Threat: A Christian Perspective, Wild Goose Publications, Glasgow, 1996.

Swanson, J., Renew, D., and Wilkinson, N., Power Lines and Health, *Physics World* November 1996, p. 29.

van der Laan, N., Struggle for life in Russia's Poison City, *Daily Telegraph* 2 May, 1997.

Ward, B. and Dubois, R., *Only One Earth*, Pelican Books, London, 1972.

7

ENERGY SCENARIOS

7.1 Sustainable Development

In the last four decades it has been increasingly recognised that we are consuming the resources of the earth so rapidly that some of them will be practically exhausted in the not too distant future. The most notable example is oil, that is now being used on a large scale and will last only about fifty years at the present rate of consumption. At the same time, we are polluting the earth on such a scale that already in many countries forest and lakes are dying.

Such considerations show that our present life-style is not sustainable; we cannot go on like this for ever — sooner or later it will break down. We are failing in our responsibilities to future generations: we are not ensuring that we leave the earth in a state fit for them to live in.

Fortunately, the earth has considerable powers of recuperation. If pollution is controlled, the lakes come to life again and the forests regenerate. Furthermore, the advances of technology make new materials available, so we do not have to ensure that future generations can live in just the same way as we do. For many applications, plastics has replaced steel, and uranium can replace oil as our main source of energy. Our concept of sustainability is not static but dynamic.

Viewed in this way, sustainability becomes a practicable aim. It is of course necessary to curb pollution, and to re-use waste material or dispose of it in a safe manner. Some types of waste material can be burnt or degrades naturally, but others remain, and these require special attention. The earth has vast quantities of many of the most important materials, such as water and sand, and so there is no need to be concerned about future supplies. Others

139

are required in such small quantities that even quite small reserves will last almost indefinitely. The important quantity is the ratio of the total reserves to the annual use that gives the number of years it will last before being exhausted. Materials for which this number is small need to be conserved and the possibility of substitutes studied.

In the context of energy supplies, these considerations support the use of renewable sources whenever possible since, apart from the materials used in the construction of dams or wind turbines, they do not consume any resources. Unfortunately it is not possible to obtain more than a small fraction of the energy we need in this way. Since we have to use fuels that are extracted from the earth, these should be examined for the pollution caused by using them and also whether they are being used up at a rate that jeopardises other important applications.

Several scenarios have been proposed to ensure our future energy supplies. At one extreme is an all-out development of all energy sources both new and old, including nuclear, in order to bring the developing countries up to the level of energy consumption, and hence standard of living, of the developed countries. At the other extreme is the green scenario that relies on the renewable energy sources and excludes nuclear and possibly coal power as well. In between is a range of options that most countries adopt in one way or another.

In 1975, Prof. B. T. Feld gave the goals for the year 2020 that are listed in Table 7.1.

Table 7.1. Future Energy Scenarios (Feld, 1975).

Model	kWh per cap	Total in 10^6 MW	Factor of increase since 1970
Optimistic	10	100	14
Prudent	3	20	4.3
Conservative	2	20	2.8

Table 7.2. Growth rates needed to achieve the levels of energy in Table 7.1.

Model	Years at 5% Growth	Doubling Time (years)	Annual Growth Rate
Optimistic	88	8	9
Prudent	65	11	6.5
Conservative	55	13	5.5

The capital costs of these scenarios are given in Table 7.3.

Table 7.3. Capital Costs of Energy Scenarios.

Model	Total Cost (10^6 B)	Annual Cost 50 yrs (10^6)	5% (10^6) Growth
Optimistic	79	1600	900
Prudent	23	460	350
Conservative	14	280	250

Initially, in the nineteen fifties, there was much enthusiasm for nuclear power, and it was confidently expected that it would soon replace the other major sources. Thus in 1970, when its contribution to UK's electricity supply was 13%, it was expected to rise to 65% by 1985 and possibly over 90% by 2000. In reality, it reached 37% by 1997. The world nuclear energy generation was projected to be 600 GWe in 1985, rising to 2000 GWe by 2000. In reality, it reached 350 GWe by 1996.

If the present nuclear production were held to 400 GWe for 100 years, this would generate 70 TWy of thermal energy, compared with 154 TWy for oil and 130 TWy for gas (Hafele, 1990). This will require about 7 million tons of uranium, similar to the currently estimated world supply. These figures show that if nuclear power is to make a dominant contribution to world energy needs the development of fast reactors and their widespread use is essential.

In order to achieve a sustainable society it is desirable to ensure that the price of any commodity, including energy and manufactured goods, includes the cost of removing or disposing all waste products, and preventing pollution of the earth so that the environment is not affected in any way that is likely to have an adverse effect on our lives.

7.2 Low Energy Scenarios

Our modern technological society is certainly polluting the earth on an unprecedented scale. Huge areas are covered by sprawling cities, motorways and industrial sites. The air we breathe is increasingly polluted, and the acid rain is killing forests and lakes. We are rightly appalled by all this, and certainly something must be done, or eventually we will all die.

It is easy to put the blame for this on the technological society as such, and to say that we must return to a simpler lifestyle. We can build our homes more simply and grow our own food without pesticides or factory farming. We can obtain our modest needs by burning wood and dried plants, and generate some electricity by windmills and solar panels.

The motivation behind all this is good. There is no doubt that we who live in highly industrialised societies could, and indeed must, moderate our life styles. We can easily, without undue discomfort, use far less energy simply by turning down the heating in winter and the air conditioning in summer, by walking short distances instead of using the car, by using public transport wherever possible and by foregoing unnecessary journeys. Elimination of such waste of energy is needed not only to preserve the earth, but to make more energy available to those millions of people who lack the basic necessities of life.

Even if all this is done, we still need very large amounts of energy to preserve a tolerable lifestyle. It is neither practicable nor politically acceptable to return to the lifestyle of even a century ago. The scientific and technological developments of the last two centuries have brought immense benefits that cannot be abandoned. Our medical care, education, transport and communications, to mention just a few of the benefits, are undoubted gains. The mass production of goods has brought cheap clothing, household goods and food to millions of people. All this requires energy on a scale that cannot be met by windmills and solar panels. Anyone who doubts this should examine in detail the quantitative estimates of the potentialities of these and similar energy sources that have been discussed in the preceding chapters.

An argument that is frequently deployed is to say that since a certain energy source can produce say X megawatts, then to supply the needs of a country that requires Y megawatts all we have to do is to build Y/X of these sources. For example, a windmill can generate 1 MW, so to meet the needs of a country consuming 500,000 MW, all we need to do is to build 500,000 windmills. The difficulty with this argument is that the various factors concerned with the manufacture and erection of these windmills do not scale in the same way. While there are some economies from mass-production, the huge demand for construction material may well increase their cost. Then there is the problem of finding suitable sites, together with the necessary permissions. This becomes increasingly difficult as the number increases. Thus when estimating the cost of a large number of power plants it is not just a matter of multiplying the

cost of a few. In addition there are problems of the availability of the
energy: what to do when the wind does not blow. Similar problems would
be encountered for other energy sources.

It is good to love the green countryside and clean air, and these must
be preserved as far as possible. We have other needs as well, and we cannot
abandon the technological society. Our needs and our desires are incompatible,
and so we are inevitably faced by hard choices. Shall we build a dam to provide
water and power for a city, but will also destroy a beautiful mountain valley,
and ruin the lives of the people who have lived there for centuries? Shall we
build a new airport or motorway, knowing that it will destroy valuable land
and severely affect the lifestyles of those living nearby? These and hundreds
of similar questions are inevitable in our modern world, and they need to
be tackled with great care and sensitivity. It is no longer tolerable to allow
industries to despoil the landscape and pollute the air and the rivers, driven
solely by the motive to maximise their profits. Stringent regulations must
be devised and enforced to reduce pollution to a minimum without at the
same time destroying the economic viability of the industry. This may not be
practicable for one country acting alone, since pollution control is expensive
and thus puts an industry at a disadvantage compared with its competitors in
other countries who may not be subject to the same restrictions. It can only
be tackled at the international level, perhaps through the United Nations.

The low energy scenario in its extreme form must therefore be judged
impracticable, but it is motivated by genuine concerns that must be taken
seriously in order to preserve the earth.

It is particularly important to recognise, once and for all, that we cannot
obtain our energy needs solely from windmills and solar panels, though they
deserve to be used as far as practicable. Large power stations are absolutely
necessary to supply even our minimum energy needs. The question then arises
as to which type of power station does least damage to the environment. Such
questions must be faced by proponents of low energy scenarios.

In order to stabilise the emission of carbon dioxide by the middle of the
next century we need to replace 2000 fossil fuel power stations in the next
forty years, equivalent to a rate of about one per week. Can we find 500 sq km
each week to install 4000 windmills? Or perhaps we could cover 10 sq km of
desert each week with solar panels? Tidal power can produce large amounts
of energy, but can we find a new Severn estuary and build a barrage costing
£9B every five weeks? The same sort of question could be asked about nuclear

power. The answer is that in the peak period of nuclear reactor construction in the 1980's the average rate of construction was 23 per year, with a peak of 43 in 1983. A construction rate of one per week is thus quite practicable.

7.3   High Energy Scenarios

The world population is growing rapidly, doubling every 35 years, and the demand for energy is increasing even more rapidly, as living standards rise. In addition there are billions of people living in poverty, and raising their living standards requires a further increase of energy production by at least a factor of five. We want to save the remaining oil for more specialised applications, and coal is far too polluting, so we must rely on nuclear to provide all the energy we need. At first we use thermal reactors and then, as the price of uranium rises, we can bring in the fast reactors which will provide our energy needs until the fusion reactors become available half-way through the next century. We have the technology to do all this, so it can be done.

It is obvious, however, that if the world population goes on increasing exponentially without limit this scenario, or indeed any other scenario, will become impossible. Sooner or later the population must be stabilised and the urgent question is whether it does so in a controlled and planned way or by a series of catastrophes. As we have seen in Chap. 1 there are good reasons to hope that the world population will level off at about twelve billion, and then the high-energy scenario becomes technically possible.

We then have to ask what is the likely effect of this high energy scenario on the environment? It must first be noted that the nuclear power stations needed will take many years to build, and in the meantime it is necessary to rely on the existing fossil fuel power stations. Under normal operating conditions, nuclear power stations cause very little atmospheric pollution, whereas the only practicable alternative, coal, is seriously polluting. Thus as the nuclear power stations are built, they replace coal power stations, reducing the pollution. However even if we replace all fossil fuel power stations by nuclear power stations we have to consider the environmental effects of the factories making these power stations, and all the associated industries. The extra energy that will become available will make it possible to build more factories to produce the goods needed by people in the poorer countries. It is implicit in this scenario that the developing countries will achieve their goal of raising their living standards to those in the presently developed countries. This is of course a matter of extreme difficulty that is unlikely to be achieved for a very long time. These problems, which are primarily political, are not discussed here, but only the question of supplying the energy they need.

In the early days of nuclear power there was great optimism that here was the energy source that would provide for all our needs. In an article in 1969 the demand for electricity in the year 2000 was estimated to be 10,000 GWe, requiring the installation of new power stations at the rate of 500 GWe per year, that is a new 10,000 MW power station every week. In that period world energy needs were expected to increase by a factor of four, half due to population increase and half due to increasing energy consumption per capita. Increasing the size of nuclear power stations was expected to halve the cost of electricity. The development of nuclear power was expected to reduce the demand for coal and oil so that their prices will fall, greatly benefiting the people in the poorer countries, especially those living in remote areas not yet connected to the national electricity grid.

This optimism for nuclear power, so strong in the 1960's, has now vanished. While in some respects the high energy scenario is technically possible, it is certainly politically impossible. The peak period for the construction of nuclear power stations was in the 1980's, and it is now at a much lower level. The reasons for this rejection of nuclear power have already been mentioned — the association with nuclear weapons and the danger of proliferation, the fear of nuclear radiations, and perhaps most of all the horrific accident at Chernobyl. All these reasons, it should be noted, are more emotional than rational. In addition the activities of some environmentalists who have maintained continuous opposition to nuclear power, have persuaded large numbers of people that nuclear power is not an acceptable energy source. This is so deeply rooted that it will influence political decisions for a long time to come.

Thus, irrespective of the arguments for and against, the high energy scenario is not a viable option.

7.4 Intermediate scenarios

Since the low energy and the high energy scenarios discussed in the last two sections are both impracticable, it remains to find a way that will combine the merits of both, while avoiding as far as possible their difficulties.

First of all, as already emphasised, we must moderate our lifestyles so as to use less energy and thus reduce the harm to the environment. The second urgent need is to examine all energy sources as impartially as possible, evaluating for each of them their capacity, cost, reliability, safety and effects on the environment. Whenever possible this should be expressed numerically. Those who genuinely care for the environment must realise the seriousness of the pollution due to coal power stations and if this is coupled to a willingness to analyse energy needs quantitatively it should be possible to devise an energy policy to maximise the benefits and minimise the hazards and disadvantages.

It is not possible to articulate a general solution because it depends on the overall situation in each country or region. Hydropower is very useful in mountainous countries, geothermal and tidal in special conditions, and the other renewable energy sources are particularly valuable for some relatively small-scale applications such as providing power in remote regions where the ordinary supply is not economically practicable. Large-scale energy generation can be provided by nuclear though it will take time to replace the existing coal and oil power stations. It is essential that the new nuclear power stations are designed and maintained to the highest safety standards, and that full information on their operation is easily available.

7.5 Conclusion

This survey of the problems of energy and the environment shows that they are part of the much larger problem of the whole future of mankind. Whatever we do, it is inevitable that the population of the world will go on increasing, and the best we can hope for is that it will eventually stabilise at about twelve billion by the middle of the next century.

To provide the energy needed by all these people, and to increase the standard of living of the poorer ones, will certainly require a greatly expanded power programme. Since the world's oil supplies will soon reach their maximum, we cannot (without drastic action) hope to keep world energy production even at its present level. To avoid a drop in living standards it is necessary to plan now.

There is a wide range of options. If we do little or nothing, more and more people will find themselves cold and hungry in a few years' time. To avoid this, we must make plans for the future, and if they are to have any hope of success they must be based on accurate knowledge of the capacities of the various power sources, their practicabilities, their costs, their reliabilities, their safeties and their effects on the environment.

Once we have this knowledge, we can think about the sort of world we want to live in, and make our plans accordingly. We must use existing power sources, since it takes a long time to develop new ones. Of those that are available, we have already seen that although other sources are useful on a relatively small scale, only coal and nuclear have the capacity to supply the bulk of our needs.

We could decide to do without nuclear power and increase the production of coal to supply most of our energy needs. We would then have to accept the increased costs in mining deaths, and the pollution of the atmosphere as we mined more and more coal. We could cut down the pollution by installing more equipment to clean the effluent gases, but this would greatly increase the cost. We would have to decide whether we would be willing to pay the high cost of relatively clean air and reduced pollution. The poorer countries would come off rather badly, because few of them have large reserves of coal, and so they would have to import it from other countries. This would be relatively expensive, since coal is not a convenient fuel to transport large distances, especially to countries far from the sea.

Another possibility would be to build wind and solar generators and hope that they will provide enough energy to replace the oil, without the need to increase coal production. This will greatly increase the cost, and will be subject to unpredictable breakdowns. Since wind and solar power are relatively hazardous and also have severe effects on the environment, this must be added to the cost of this option. The high cost would make it difficult to provide aid to the poorer countries. Not only would the developed countries be poorer themselves and so have less to spare, but whatever we could afford would provide less energy than the cheaper sources. Reliance on wind and solar power condemns the poorer countries to be without the energy they need.

These two options, or combinations of the two, would avoid the use of nuclear power. It is unlikely that people would be happy with them, once they understand the full implications.

Nuclear power is thus inevitably an important contributor to our future energy needs. Since it already provides almost half the electricity of several West European countries, there is no doubt that it can deliver energy on the scale needed. Hundreds of nuclear power stations have been operating safely for tens of years, so it is a thoroughly well-tested power source. The possible hazards are well understood, and the method of dealing with nuclear waste is well understood. This new technology can proved the bulk of the world energy needs more safely and economically than the alternatives, and with less damage to the environment.

Whatever course is taken, the richer countries will manage somehow or another, though probably not without economic crises. Some countries, such as Britain and the United States, have large supplies of coal and oil, and so need only develop nuclear power in the context of a balanced energy programme. Other countries, such as Japan and France, are not so fortunate and so if they are not to depend on imported fuel they have no alternative but to rely increasingly on nuclear power. Countries that do this soon enjoy the economic advantages of cheaper electric power.

Through the ages mankind has used in succession wood, coal and oil. As each new energy source has become available more sophisticated uses have been found for the earlier fuels, so that it becomes wasteful to burn them. We are now facing another crisis of transition, as falling oil supplies fail to meet the needs of a rising population. We have found a new source of energy in the nucleus of the atom, and this is not only able to provide the energy we need, but is cheaper, safer and cleaner than coal and oil. This new energy comes from uranium, which is otherwise practically useless, and by obtaining energy in this way we conserve the wood and coal and oil so that they are available to satisfy more specialised needs.

The possibility of further nuclear accidents must be assessed as objectively as possible. The science of risk analysis is still in its infancy, but it is infinitely preferable to allowing our future energy policy to remain at the mercy of political decisions. Whatever we do, we cannot entirely eliminate the possibility of an unexpected event or of human negligence and folly. Despite all our essential efforts to improve safety, there will always be accidents and disasters: ships will sink, dams burst, oil rigs capsize or catch fire, and mining galleries collapse. It is one of the inescapable conditions of living in a technological society. We must do all we can to minimise accidents, but must learn to live with them when they occur. The only alternative is to turn our back on technology, and then we would rather rapidly sink back to a primitive level of existence.

The solution of these problems is not made any easier by the political climate in many countries. The result of a long campaign against nuclear power, reinforced by media more concerned with sensation than truth, is that vital decisions are often decided not by objective scientific evidence but by political considerations. Thus although careful studies showed that it is safe to dispose of low level radioactive wastes in the sea, it is politically unacceptable.

Another example is provided by the stringent rules imposed on the nuclear industry concerning radioactive emissions that far exceed those imposed on other industries. This implies that lives would be saved if uniform pollution controls were applied to all industries. The problems facing politicians are severe enough, without them being forced by political pressure to take decisions that they know are opposed to our long-term benefit.

It has been very reasonably suggested that a carbon tax should be imposed on all methods of power generation that emit carbon gases into the atmosphere. The rate of tax would be based on estimates of the damage to health and to the environment due to these emissions. This would immediately increase the price of coal power by about a factor of four, making it far less economically acceptable. This shows how much we are really paying for our continued reliance on coal power stations.

Since nuclear power is dependent on uranium, it is important to know how long the supplies will last. In recent years the discovery of new deposits and the slowing down of the nuclear programme in some countries has led to a large reduction in the price of uranium. As soon as the demand for uranium starts rising the price will go up and this will render economic many mines that were not profitable before. It will be a long time before there is a world shortage of uranium. Before this happens, the fast reactors will have reached the stage of commercial development. It has already been estimated, for example, that in Britain the depleted uranium piling up around nuclear power stations has an energy content greater than that of all the oil in the North Sea. This is illustrated by Fig. 7.1, which shows the world fuel reserves of coal, oil, gas and

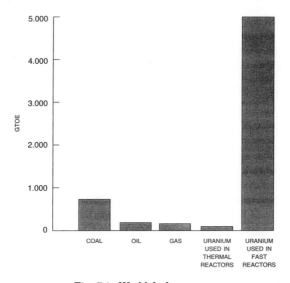

Fig. 7.1. World fuel resources.

uranium if used in thermal and in fast reactors. The latter is several times greater than all the rest put together.

There will therefore be enough uranium to last until well into the next century, and by then we may hope that the problems of fusion energy will have been solved and fusion power reactors will have become a practical possibility. Scientists and technologists have provided the means of satisfying world energy needs in a way that has little adverse effect on the environment. It is the responsibility of Governments and statesmen to ensure that the right decisions are taken to bring this about. This will be possible only if they have the support of an informed public opinion. At the present time, ill-informed public opinion often forces Governments to adopt policies that they know to be undesirable, thus making it more difficult to provide much-needed energy in an economical and environmentally friendly way.

8

POLITICAL ASPECTS OF
THE ENERGY CRISIS

8.1 Introduction

The development of energy sources and their effects on the environment are inevitably political problems. The choice between energy sources and what to do about pollution have to be decided by Governments. Strong passions are easily aroused, pressure groups are formed, and truth is the victim. This has had a great influence on the development of energy sources, especially nuclear power, and this is described in Sec. 8.2.

Since most people are innumerate, the arguments are qualitative and frequently emotional. So many questions concerning energy can only be settled with the help of some numerical estimates, and these can only be provided by those with specialised knowledge. Scientists thus have a responsibility to contribute to the public discussions in order to ensure that they are based on reliable information. The responsibility of scientists is discussed in Sec. 8.3.

One might expect that the contributions of scientists to the energy debate would be generally welcomed, but this is not the case. Most people seem to have minds that are already made up, and so they do not want to hear from scientists who might well have different views. Scientists are regarded as unreliable allies, because we are liable to change our minds in the light of new evidence. Some examples of unsuccessful attempts to inject factual realism into the energy debate are recounted in Sec. 8.4.

Much of the public debate in conducted in the mass media, particularly the Press and television. Very often this is extremely irresponsible, and some

examples are described in Sec. 8.5. The media is largely responsible for forming public opinion and the results of decades of misinformation are shown in
Sec. 8.6. The complexities of modern technology are so great that disasters
can occur because the responsibility for action is spread too widely. It is then
easy for each individual to assume that someone else is responsible, particularly if strong political pressures are being applied. This danger is discussed
in Sec. 8.7.

8.2 Politics

As soon as Fermi achieved a self-sustaining chain reaction with his pile of
graphite and uranium, those present knew that an immensely powerful new
energy source had been found. The next few years saw the secret development
of the atomic bomb, and its use to end the war with Japan. In the immediate
post-war years, many of those who participated in the atomic bomb work at Los
Alamos enthusiastically developed reactors to generate electricity for peaceful
uses. With a few years civil power reactors were operating, first in Britain,
and then in the United States and several other countries. The media were
full of optimistic visions of the future to be made possible by the new energy
source.

In many respects it was indeed a success story. The new nuclear power
stations proved themselves to be reliable and safe, and their economical aspects
compared reasonably well with those of coal power stations. Nuclear power
stations were more costly to build, but their operating costs were lower.

Gradually, however, the mood changed. The accident at Three Mile Island
severely shook the public confidence. Fears of unknown radiations, and memories of the devastation of Hiroshima and Nagasaki came to the fore, and were
fanned by media stories of thousands of people who would die because of the
accident. Although devoid of foundation such fears, once released, could not
be put back in the bottle. The nuclear industry had neglected to ensure that
people were properly informed about the realities of the nuclear age, and the
reports of the Three Mile Island accident showed deficiencies in the design and
operation of the reactor.

The reaction against nuclear power was strongly supported by political
groups who wanted to reduce the power of the West. They stirred up opposition, making full use of the fears of nuclear radiation and the memories of
the atomic bombs. Many politicians urged that legislation be introduced to
impose a strict safety review before permission to build a nuclear power station

was granted. Up to a point this is indeed necessary, but eventually the legal requirements became so lengthy and cumbersome and they increased the cost to such an extent that Electrical Power Companies found that it was no longer economical to build nuclear power stations. Thus, although by that time the United States nuclear power programme was by far the largest in the world, many nuclear power projects were cancelled and no additional stations were built.

The United States has other sources of energy, in particular immense reserves of coal. It was a pioneer in oil production, but now has to import oil. Thus it could survive without nuclear power stations, but at the cost of increased pollution. Other countries were not so fortunate. France has no oil and little suitable coal, and realised that it is politically undesirable to rely on imports. So they decided to develop nuclear power, and now about 80% of the electrical power generated in France comes from nuclear power stations. Britain has large reserves of coal and oil, and so the development of nuclear power is not imperative. Many other countries in Western Europe have large nuclear power programmes, and nuclear has now overtaken coal in the generation of electricity in that area. Nuclear power is also popular with the tiger economies of South-East Asia, particularly Korea and Taiwan.

Whether a country develops nuclear power is thus partly a matter of need, depending on its indigenous supplies of coal and oil, and partly a matter of politics. In countries where the alternatives are practicable, those opposed to nuclear power can exert sufficient political influence to slow down the use of nuclear power, as for example in Britain and the USA, or to stop it altogether, as in Germany. Where this happens, coal or oil power stations are used instead of nuclear, causing much greater pollution. This brings with it increased health hazards, acid rain and more carbon dioxide in the atmosphere. This is the legacy of the anti-nuclear movement.

The experience of Germany shows very dramatically the effects of politics on the development of nuclear power. For some years the Green parties held the balance of power, and they used their influence to stop all nuclear development. As already mentioned, the completed prototype fast reactor at Kalkar was never allowed to operate. The prototype high temperature reactor at Schmehausen was abandoned after operating for only a year. The nuclear-powered ship Otto Hahn was decommissioned after two demonstration voyages. This is not only an appalling waste of money and resources, but is a soul-destroying experience for the thousands of scientists and engineers who have spent most of their lives bringing their project to fruition. With such examples before them, there is scant incentive for the next generation of young graduates to devote their lives to the development of advanced technologies.

Another area where political considerations cause confusion and often serious harm is the imposition of unrealistic radiation dose limitations. As an example, the National Radiological Protection Board, on the basis of a limit to a whole body dose of 5 mSv set a limit of 3600 Bq/litre on milk. The European legislation limit, which by law now overrides the NRPB figure, is 1000 Bq/litre. This reduction was made for political and presentational considerations and is likely to confirm irrational fears and lead to the waste of milk that can be drunk quite safely.

8.3   The Responsibility of Scientists

The discovery of the nuclear chain reaction has the most momentous consequences for mankind. It made possible the atomic bomb, that has changed the nature of international confrontations, and it has provided a new source of energy for peaceful purposes. Scientists are responsible for initiating these changes, but what are our continuing responsibilities? I think that our responsibility is to see that the public debates on energy questions are conducted with due regard for the essential scientific facts, so far as these are known. Certainly scientists are also entitled to participate in the subsequent debate, but on political and other questions their views are hardly more worthy of attention than those of any other concerned member of the public. But as a source or guardian of balanced and accurate factual information the scientist has a unique and irreplaceable role.

In some circumstances scientists have a very heavy responsibility, when because of their specialised knowledge they alone can foresee the consequences of a scientific discovery that will strongly affect world affairs. A striking example of this is provided by the calculations made in Birmingham in 1939 by Rudolf Peierls and Otto Frisch of the amount of fissile material necessary to make an atomic bomb. If this had turned out to be several tons, then it would have been difficult to make it into a practicable weapon. They found, however, that a few kilograms are sufficient, and immediately realised the consequences. They wrote a detailed memorandum and sent it to the British Government, which then took the necessary action.

These responsibilities fall mainly on university scientists, because only we have both the knowledge and in many cases the freedom to speak. This is not, however, the case in all countries. In totalitarian countries scientists have little freedom of speech, but even in countries usually considered democratic it requires great courage to speak out against the general consensus. Such action may adversely affect one's ability to find good positions for one's students, and can also tarnish one's scientific reputation; sound scholars only talk about their speciality. Furthermore, many of the most well-informed scientists are excluded from the debate by commercial or Governmental security, and in any case are often seen, usually unjustly, simply as spokesmen for their employers. The net result of all this is that relatively few scientists are in a position to speak with some authority on matters of vital concern to society as a whole.

During the hectic years at Los Alamos when the bomb was being developed the scientists were so intent on their task that there was virtually no discussion of its social implications. After Hiroshima and Nagasaki there was an outburst of words, feelings, emotions and expressions of guilt. The scientists realised that they must educate the public about the new force that had entered world history through the work of their hands. Why should they do this? The question is answered by Laura Fermi: "Unless the public come to understand some of the basic scientific, technological and political aspects of the bomb, the Americans will not be able to make well-reasoned decisions. Our husbands went ahead and gave lectures in Santa Fe and Albuquerque and they organised the Association of Los Alamos Scientists. They drafted statements, they wrote articles".

In the late forties there was very little public understanding of nuclear power: everyone had heard about the atomic bomb, but the basic physics underlying nuclear power and the nature of nuclear radiations were not understood at all. The scientists who had participated in the wartime developments, and others who graduated since that time, realised that there was a great and urgent work of public education to be tackled. They formed the Atomic Scientists' Association in Britain, and the Federation of Atomic Scientists in the United States, both devoted to the task of educating people for the nuclear age. They published periodicals, organised exhibitions, and lectured up and down the country. They were everywhere listened to with close attention, and soon more and more people began to understand the age they were about to enter.

Their main concern at that time was to warn against the dangers of nuclear war, and to make it clear that the advent of nuclear weapons had completely changed the nature of war. Their second concern was to show how the same power could be use for peaceful purposes. This included not only nuclear power, but also the wide variety of medical applications of nuclear radiations, as well as the use of isotopes in industry and agriculture.

Most of the work of the Atomic Scientists' Association concerned the dissemination of factual information, but occasionally there were more sensitive decisions to take. One of these occurred soon after the fifteen megaton hydrogen bomb test at the Bikini atoll on 1st March 1954. Soon after, the BBC made a television programme on the hydrogen bomb during which the physics of the bomb and its military, political and ethical aspects were debated by the Archbishop of York, Lord (Bertrand) Russell and Professor Rotblat, Professor of Physics at the Medical School of St Bartholomew's Hospital in London. Professor Rotblat described the two-stage mechanism of the bomb: the fusion of heavy hydrogen isotopes that provides the main force of the explosion, the reaction being detonated by a fission bomb. The explosive power of the hydrogen bomb is about a thousand times that of the Hiroshima bomb but, since the fusion reaction produces very little radioactivity, the radioactive contamination is hardly increased at all.

The radioactive debris from the test explosion fell over a wide area, and some of it fell on the Japanese fishing boat Fukuryu Maru. The crew saw the explosion and later on noticed a whitish powder that fell on the boat. About three days later they found that those parts of the skin that had touched the powder became dark red and began to swell up like an ordinary burn. After the boat had returned to port, Professor Yasushi Nishiwaki, a professor of radiation biophysics at Osaka City University School of Medicine, read a report of this in the newspaper and was then asked by the Public Health Department of Osaka to examine some of the tuna fish brought back by the Fukuryu Maru. He found that they were highly radioactive, and was able to establish the presence of a range of fission products in the dust he found on the boat itself.

Professor Nishiwaki wrote a detailed article on his findings that was published in the *Atomic Scientists' Journal* for November 1954. These were analysed by Professor Rotblat, who found to his surprise that "fission accounts for most of the energy released by the hydrogen bomb". This was contrary to the general belief at the time that a fission bomb simply served as a detonator to initiate the much more powerful hydrogen reaction. Professor Rotblat then realised that the mechanism of the bomb was rather more complicated: there was an additional third stage composed of a shell of uranium 238. This shell serves the double purpose of holding the reacting mass together for a short time longer, thus increasing the explosive power, and far more importantly boosting the power still further by the extra fissions caused in the uranium 238 by the fast neutrons from the hydrogen reaction. Since the energy per fission is about 200 MeV, and those from the three main fusion reactions are 4.0, 3.3 and 17.7 MeV, it is easy to see that most of the explosive power from such a device comes from fission and not from fusion.

This was a startling conclusion, and shows how an academic scientist, by reasoning from published data, can reach conclusions about matters that are considered most secret by the authorities. Professor Rotblat was concerned that the account of the mechanism of the bomb that he gave in the BBC television programme was incorrect, so he wrote an article on his work that was published in the *Atomic Scientists' Journal* for March 1955. This was of course a sensational disclosure that received wide publicity. The news that the hydrogen bomb was extremely dirty from the radioactive point of view was not popular with the authorities on either side of the Atlantic, and some pressure was exerted to prevent publication. As editor at the time, I do not recall having any hesitation about publishing this article; it seemed to be our duty to make this knowledge generally available.

Lord Russell became increasingly concerned about the dangers of thermonuclear wars, and especially the long-term effects of radioactive fallout, and expressed his concern in a broadcast on 23 December 1954. He presented the choice, "stark and dreadful and inescapable: Shall we put an end to the human race; or shall mankind renounce war?" He ended; "I appeal, as a human being to human beings. If you can do so, the way lies open to a new Paradise; if you cannot, nothing lies before you but universal death". This appeal received wide publicity and he was urged to translate it into action. He drafted a statement and sent it to Einstein and other eminent scientists for their signatures. Flying from Rome to Paris, Russell learned from an announcement by the captain of Einstein's death. This was a shattering blow to his hopes, but when he arrived at his hotel in Paris he found a letter with Einstein's signature endorsing the statement, and all was well.

This work of the scientists continued for some years, and gradually the situation changed so that they felt their responsibilities to be less pressing. By then there had grown up a generation of science writers, well-trained and articulate, who progressively took over the role of informing the public. Many books were published by those involved in the wartime project so that detailed and accurate information became readily available. Many scientists, with some reason, felt that their work was essentially done, and so they could henceforth concentrate on their academic work. Others thought that the scientists still had urgent responsibilities. The result was that in Britain the Atomic Scientists' Association was disbanded, and many of its members transferred their energies to the Pugwash Movement that had been established as a result of the Russell-Einstein manifesto.

The motivation behind the Pugwash Movement is that, in the words of a document issued on the occasion of its twenty-fifth anniversary, "An ever-growing number of scientists now realise that they have to share the responsibility of Governments to utilise knowledge for constructive purposes, so that beyond the interests of individual groups and countries the achievements of science and technology shall benefit the welfare of mankind as a whole and not contribute to its detriment".

The Pugwash Movement was originally concerned mainly with the problems of disarmament, but over the years its concern has widened to include the whole field of science and public affairs. Discussions of these questions have continued over the last thirty years and in that time the situation has changed radically. The number of books on nuclear and energy problems have multiplied, news items are frequently published in daily papers, colour supplements and magazines and broadcast by television. If however you were to ask a well-informed scientist to assess the total impact of this material he would be obliged to say that it presents a gravely distorted view of the reality.

Forty years ago the scientists usually had a rather receptive audience. Now, we are caught up in a maelstrom of politics, and our task has become almost impossible. Our main concern, as scientists, is to present the facts as clearly and accurately as possible, together with the implications. Unless the basic facts are reasonably well known, there is no hope of making wise decisions. The scientists as such are neither pro-nuclear nor anti-nuclear, they simply want to present the facts as best they can. But however much scientists try to be objective and impartial, they are easily seen as partisan. As the views expressed in the media swing from one extreme to the other, the scientist, simply by striving to tell the truth, is at any one time almost always pulling against the media. In the days of uncritical enthusiasm for nuclear power, scientists who ventured a word of caution would be branded as anti-nuclear. Now, with the media predominantly anti-nuclear, they inevitably appear to be pro-nuclear.

The reasons for the worsening of the energy debate are complicated, and include political and psychological considerations. There is the association between nuclear power and nuclear weapons, and the fear of unseen nuclear radiations that are known to cause cancer. These fears have been amplified over the years by left-wing pressure groups financed by the Soviet Union in a sustained attempt to de-stabilise the economies of the West. They know very well that only scientists can provide the factual information that will expose the falsity of their propaganda, so every effort is made to discredit them by attacking the objectivity of science and by saying that their views are due to their support of the current capitalist-imperialist establishment. A decade or so ago most of their energies were concentrated on the campaign for nuclear disarmament (again with the purpose of weakening the West) but with the collapse of the Soviet Union their energies have been directed to environmental problems, but still carrying with them their hatred of anything nuclear.

The result of all this is that the scientists have lost the battle to convey the facts about nuclear power to the society in which we live. The way this happened in the United States has been well described by Professor Bernard Cohen:

"First let us consider the cast of characters in the battle. The two sides are of an entirely different ilk. One of the main interests in life for a typical anti-nuclear activist is political battling, while the vast majority of nuclear scientists have no inclination or interest in political battling, and even if they did they have little native ability or educational preparation for it .... While the former was making political contacts and developing know-how in securing media co-operation the latter was absorbed in laboratory or field problems with no thought of politics or media involvement. At this juncture the former went out looking for a new battle to fight and decided to attack the latter; it was like a lion attacking a lamb.

Nuclear scientists had long agonised over such questions as what safety measures were needed in power plants, and what health impacts their radio-activity releases might cause. All the arguments were published for anyone to see. It took little effort for the anti-nuclear activists to collect, organise selectively, and distort this information into ammunition for their battle. Anyone experienced in debate and in political battles is well prepared to do that. When they charged into battle wildly firing this ammunition, the nuclear scientists first laughed at the *naiveté* of the charges, but they didn't laugh for long. They could easily explain the invalidity of the attacks by scientific and technical arguments, but no one would listen to them. The phoney charges of the attackers dressed up with their considerable skills in presentation sounded much better to the media and others with no scientific knowledge or experience. When people wanted to hear from the scientists, the attackers supplied their own — there are always a few available to present any point of view, and who was to know that they represented only a very tiny minority of the scientific community. The anti-nuclear activists never even let it be made clear who they were and who they were attacking. The battle was *not* billed as a bunch of scientifically illiterate political activists attacking the community of nuclear scientists, which is the true situation. It was rather represented as 'environmentalists' — what a good, sweet and pure connotation that name carries — attacking big business interests (the nuclear industry) which were trying to make money at the expense of the public's health and safety.

The rout was rapid and complete. In fact the nuclear scientists were never even allowed on the battlefield. The battlefield here was the media which alone have the power to influence public opinion. The media establishment swallowed the attackers' story hook, line and sinker, becoming their allies. They freely and continually gave exposure to the anti-nuclear activists but never gave the nuclear scientists a chance. With constant exposure to this one-sided propaganda, the public was slowly but surely won over. The public was driven insane with fear of radiation; it became convinced of the utterly and demonstrably false notion that nuclear power was more likely to kill them than such well-known killers as motor vehicle accidents, cigarette smoking and alcohol; that burying nuclear waste, actually a very simple operation, was one of the world's great unsolved problems; that, contrary to all informed sources, the Three Mile island accident was a close call to disaster and so on. Fears of everything connected with nuclear power were blown up completely out of perspective with other risks. Hitler's man, Goebbles, had shown what propaganda could do, but the nuclear scientists never believed that it could succeed against the rationalism of science; yet succeed it did. The victory of the anti-nuclear activists was complete.

The anti-nuclear activists have won the battle, and to the victors belong the spoils — the failure of nuclear science to provide the cheap and abundant energy we sorely need. That is the goal they cherished and they have achieved it. Our children and our grandchildren will be the victims of their heartless tactics".

Some illustrations of what happens when a scientist attempts to provide some factual information are given in the next section.

8.4 Some Case Histories

Over the last forty years I have made numerous efforts to provide factual information for the nuclear and energy debate, with conspicuous lack of success. As an illustration I recall an attempt to discuss the effects of nuclear radiations with a well-known politician. In a public speech, he said that if there were an accident at a proposed nuclear power station in Britain similar to that at Three Mile Island there would be thousands of deaths from cancer. This figure seemed rather high to me, so I wrote to him asking if he would kindly tell me how the figures were obtained. I added that I would be grateful to him for his permission to quote his reply.

He answered saying that he regretted that he could not find the documents that he had used to obtain these figures, but he assured me that at the time he made the statement he had the authority on which he could, as a lawyer, have relied if his statement had been questioned.

Meanwhile, I found that Dr. Martin Goldman of the Energy-Related Health Research Laboratory of the University of California had obtained a figure of 0.4 extra cases of cancer in the surrounding area due to the Three Mile Island Accident. I wrote again to the MP, quoting this figure, adding that it was inconsistent with the one he had quoted so that his statement could give rise to unnecessary public anxiety. I expressed my regret that he had been unable to justify his figures, and asked if he had any further comments to make.

In his reply he began by saying that he regarded my letter as impertinent, adding that he was quite certain that he would be able to find the source of his figures. He went on to say that if I had done a little more research I would have found the source myself. He recommended that I seek out a certain doctor, whose name he was unfortunately unable to recall, who had been making many public statements on the dangers of radiation. He added that if I was not too lazy I could easily do this, but concluded that he doubted if I would take the trouble to do so.

I therefore decided to repeat the calculations for myself. Published figures show that the accident at Three Mile Island increased the average radiation dose to the people in the surrounding area by about one millirem. Reviews by the National Academy of Sciences and the United Nations give an estimate of the cancer risk as about 100 cases per million person-rem. Thus for every million people we expect an additional tenth of a case. This figure is quite similar to that of Dr. Goldman. I wrote a further letter saying this, but received no reply.

About a week later, I read an article in *Atom* that referred to some writings of a scientist in the United States, who obtained figures for the number of extra deaths from the Three Mile Island accident rather similar to those originally quoted by the MP. It was further explained how these figures were obtained from the published data by an elementary statistical error. The actual figures showed no evidence of any effect attributable to the accident. I therefore sent a copy of the article to the politician, but again received no reply. Wishing to publish this story, I asked him for permission to quote from his letters, and received an indignant refusal. Thus I had spent a considerable amount of time with no effect whatsoever.

It is of course true that politicians are very busy people, and it is always tiresome to have one's views challenged. But the scientist knows very well that it is always important to listen to adverse criticism because he might be wrong. Indeed we should welcome informed criticism, because who in their senses wants to go on saying things that are not true? And yet it is undoubtedly the case that there are many people, including some in high places, whose minds are so firmly closed that they are unable to entertain the possibility that they are wrong, or to listen to any counter arguments. This makes it quite impossible to carry out the reasoned dialogue that is the only way to reach the truth.

Another example is provided by a leading article in a prominent daily newspaper reporting a large increase in the death rate in the United States due to the dust from Chernobyl, complete with a large picture of Death the Reaper. It was obvious that this was extremely unlikely because no such effects had been reported from Europe, when the amount of dust deposited was much larger, though still far below that likely to cause any detectable effects. The article was however supported by statistical data apparently showing a strong correlation between the amount of dust deposited and the death rate in that area.

Since the story seemed to be so unlikely, I contacted Harwell, and asked them to obtain the detailed figures for me. This took some time, and when they arrived it was clear that the figures had been obtained by statistical fudging, and that they showed no effect whatsoever. I wrote to the newspaper, but was told that it was now so long ago that everyone would have forgotten all about it. It is however more than likely that they remember the association between nuclear power and Death the Reaper.

It is greatly in the public interest that these matters should be treated as objectively as possible, taking full account of the scientific evidence. This would avoid much unnecessary anxiety, and enable the best decisions to be taken concerning our future energy supplies.

Unfortunately the treatment of such matters in the media is still far from ideal. As an example one may cite a recent advertisement by Greenpeace which appeared in several national newspapers a few months ago. This advertisement showed a photograph of a baby, described as a Kazakhstan nuclear test victim, followed by the quotation, "Hush mother do not cry. I am filled with angels". The advertisement continues: "These brave calming deathbed words of a child radiation victim may shock us. They should not surprise us". The impact of this emotion-laden photograph is somewhat changed when one learns from other sources that the child is suffering from hydrocephalus, and according to Prof. Trott, professor of radiation biology at St Bartholomew's Hospital, no case of hydrocephalus has ever been identified as liable to have been caused by radiation exposure.

The advertisement continues: "Ever since Hiroshima, children have borne the brunt of the nuclear industry's fall-out. In the womb, and as they grow, children are more vulnerable to the effects of radiation. Therefore they suffer more from radiation linked diseases such as leukaemia, foetal malformation and other genetic defects. In some ways, because of this sensitivity, they protect us. Acting as some awful early warning device. These early warning signs have been seen near to nuclear installations such as Sellafield, where plutonium is processed for nuclear weapons ... children have died. We know something is wrong". These statements about the sensitivity of children to radiation are correct, but the connections with Hiroshima and Sellafield are not.

The implications of the reprocessing plant THORP are now mentioned: "As if this were not enough, we now face the prospect of huge increases in radiation, heightened chances of accident and greater plutonium risks, all from THORP, the newly licensed nuclear reprocessing plant at Sellafield. It is our opinion that these risks are real. They beckon a world where radiation linked disease becomes an accepted part of everyday life. As if to prove the point, it can be shown from official figures that 2,000 will die because of the discharges from Sellafield over the next ten years".

This statement is untrue. The amount of radiation that will be emitted by THORP is minute, comparable in magnitude to the extra radiation experienced on a short airplane flight, or a visit to Cornwall. No one, to my knowledge, has ever cancelled a holiday in Cornwall because the natural background is two or three times the natural average. The figure of 2,000 deaths from the discharges from Sellafield over the next ten years is based on the assumption that the death rate is proportional to the dose, even for very small doses, and this is a very implausible hypothesis. It assumes, in effect, that the body is unable to recover from small doses in the same way that it cannot recover from massive doses. Making the proportionality hypothesis, using the mortality rates from massive doses and multiplying by the number of people in the population, gives a mortality figure similar to that quoted. This is obviously a totally unjustified procedure and the result has no meaning.

The advertisement concludes: "Greenpeace will go on fighting against a rationality which allows children's lives to be weighed, like so many molecules, by an industry intent on spreading radiation and the means of mass destruction around the globe. Please help us to continue the fight. To be on our side all you need is a sense of right and wrong and to refuse to be walked over by powers who deem it right to play with children's lives. In the end we will win".

To say that the nuclear industry is careless about children's lives and is intent on spreading radiation around the globe is a completely unjustified calumny on an industry that spends hundreds of millions of pounds on ensuring the highest safety standards, and has a safety record second to none.

It is very curious that an organisation with considerable resources should publish such statements. Nuclear power stations in operation make no contribution to acid rain or to the greenhouse effect, and actually emit less radioactivity than coal power stations. They are therefore making a massive contribution to preserving the cleanliness of our environment. It is difficult to understand why an organisation that claims to be working towards a clean environment is not fully supportive of the nuclear industry.

It is also difficult to see how organisations that produce advertisements such as this are contributing responsibly to the public debate about an optimum energy programme. It may be noted that this advertisement has been the subject of complaints to the Advertising Standards Authority, and after extensive expert study the complaints have been upheld. This provides an authoritative endorsement of the above remarks.

The advertisement ended with an appeal for donations (£14.50 single; £19.50 family). Doubtless many good people, their hearts touched by the picture of the dying babe and horrified by the prospect of THORP's operators deliberately multiplying such tragedies, have responded generously. Let us hope that their contributions will not be used to produce more advertisements of the same genre.

Immense harm has been done by the widespread concern over the groundless possibility that the excess cases of childhood leukaemia can be attributed to nuclear power stations and reprocessing plants. This has diverted attention and resources from attempts to find the real causes and, still more important, from studies of the environmental causes of the large number of cancer deaths, about 70,000 per year in Britain. Chemical additives to our foods, carcinogenic pollution of the atmosphere and the normal constituents of some foods are all possible causes, and research to identify them would be most valuable.

8.5 The Media

Far more poisonous and damaging than physical waste is the printed, verbal and visual waste produced by the mass media. This waste cannot be contained or buried, nor does it decay away. It spreads and proliferates, creating a climate of opinion in which any objective consideration of energy and environment becomes almost impossible.

There are indeed some science journalists attached to the more reputable papers who do their best to present objective discussions, but they are very much in the minority. Throughout much of the media, the over-riding priority is to sell as many copies as possible or to achieve the highest audience ratings. Almost everything is sacrificed to this end. Most people do not much like reading objective scientific discussions; they prefer scandal, horror and sensation, and the media are not slow in supplying it to them. In addition, there are powerful commercial and political forces that have already prejudged the issues and decided the answers. The situation is further complicated by 'nuclear consultants' who are often employed by environmental organisations and spread false information presented in scientific terms to give it a veneer of respectability.

As a trivial though typical example of journalistic overkill, the *Newcastle Journal* in 1997 had a headline: "Fear at N Plant Train Scare". What actually happened was that a train carrying an *empty* spent fuel flask from Sellafield to Hartlepool was stopped for a few minutes because an axle overheated. Anti-nuclear campaigners were quoted as saying that there could have been a tragedy if the axle had set fire to the train. Countless stories such as this have formed public opinion on nuclear power.

The way the Press treats nuclear power is hardly surprising, since it is the stated policy of the National Union of Journalists, formulated in 1980, "that nuclear power poses too great a danger to this and future generations". It would seem preferable to seek the truth in as objective a way as possible.

Most of the questions concerning the optimum energy sources can only be decided wisely after a careful analysis of all relevant factors, and scientists have an essential role in this debate. It is notorious that their participation is not welcomed; indeed they are kept out. Their contribution might tell against the prevailing political view and even if it does not they are liable to change their minds in the light of new evidence.

Thus scientists are seldom asked for their views; and if they are their words are often quoted out of context in a way that supports the media line. Sensational and wholly misleading articles are published, and if a scientist tries to correct them he is usually ignored. The result of decades of misinformation is that the public opinion on energy sources is far removed from reality.

One of the principal results of the exclusion of scientists from the energy debate is that it is innumerate. Opposing views are supported by rhetoric and emotion and illustrated by error and misrepresentation. Facts and figures are conspicuous by their absence. Since in most cases the best course of action can only be determined by comparing numerical estimates of pollution or cost or risk, this essential activity is made impossible.

Governments usually have highly-qualified scientific advisors who are able to give them objective and accurate advice, but politicians with their eyes on the next election are extremely responsive to public opinion. They are pressured to take decisions that are contrary to the best advice. In this way they remain popular, but do incalculable damage. It takes a very strong politician to take the right decision in the face of adverse public opinion.

There are genuine political questions that need widespread public discussion. The economic and technological realities define the area of realistic action, but still leave many options open. We have to balance the competing demands of safety, cost and environmental damage. In practical terms, how much are we prepared to pay for extra safety or to reduce environmental damage? These are real questions to which everyone can make a useful contribution. This debate hardly ever takes place because the technological realities that define the area of realistic debate are not known or accepted.

Environmentalists are not immune from political pressures, and many environmental organisations show the same reluctance to listen to scientists. Very frequently they instinctively support what they can the 'benign renewables', apparently unaware of the damage they do to the environment. They have the vision of a pastoral Arcadia with a few picturesque windmills to supply all our energy needs. There are, however, some encouraging signs that many environmentalists, when they actually see a wind farm in operation, are becoming aware of the environmental damage they cause.

Large power stations are rightly the concern of all who care for the environment, but if we are not to return to the stone age we cannot do without them. What must be decided is which type of power station is least damaging to the environment, assuming of course that it is also safe, reliable and economic. As we have seen, the principal alternatives are coal and nuclear. Nuclear power is usually anathema to environmentalists, to whom it is the symbol of all that is hateful in our modern technological society, and so they prefer coal power stations, which are far more damaging to the environment.

If the advocates of such policies are questioned it rapidly becomes clear that they have not done their sums. They seem not even to know what it is to study a question scientifically. At best, they have consulted some politically-motivated sociologist specialising in energy matters, yet are markedly reluctant to seek genuine scientific advice. Not only do they not know, they seem not to want to know, or even to know what knowing means.

8.6 Public Opinion

In a democratic society, public opinion is extremely powerful, and it can easily force Governments to take decisions that they know are contrary to the public interest. It is proposed, for example, to build a nuclear waste disposal facility at a certain place. Detailed scientific studies have established that this poses no danger to the public. Nevertheless, as soon as the decision is announced, a public protest group is formed to campaign against the decision. Using the familiar emotive arguments about radiation hazards, it soon gathers support. The member of Parliament for that area realises that if the Government persists in this decision, he is in danger of losing his seat at the next election, so he urges the Minister to reverse his decision. The Minister, torn between his public duty and his party loyalty, can easily give in and refuse permission for the disposal facility. Many other examples could be cited, such as the decision to stop deep sea waste disposal, in spite of scientific studies that have demonstrated its safety.

Public opinion, so powerful and influential, is very ill-informed, and its knowledge does not correspond to reality. It is strongly affected by events like the Chernobyl disaster, which turned many people against nuclear power.

Some countries, such a Sweden, voted to phase out nuclear power as soon as possible, although they later realised that there is no alternative that is not highly polluting, destructive of the environment and likely to double the price of electricity, making industries uncompetitive and increasing unemployment.

In an investigation of public opinion, 11,819 people in twelve EEC countries were asked if the risks associated with nuclear power stations are 'worthwhile', 'of no concern' or 'unacceptable'. The respective percentages changed from 43, 7 and 38 in 1984 to 29, 7 and 56 in 1986 after Chernobyl. The percentage regarding nuclear and the least polluting energy source has increased from 5.7 in 1986 to 8.5 in 1990, and many people (38%) believe that the renewables are the least polluting.

The pressure of public opinion has forced Sellafield to spend £250M to reduce radioactive discharges by an amount estimated to save one or two hypothetical statistical cancer deaths over the next 10,000 years. The Radioactive Waste Management Advisory Committee commented that such expenditure "may appear excessive". Certainly, many lives could be saved by spending such sums in other ways. For example, installing motorway crash barriers saves one real statistical life for £5,000. Here again, public opinion is killing people.

Another example of double standards is the difference between the recommended maximum dose for radon in houses (10 mS/yr) and that for radiation from nuclear reactors (1 mS/yr). This indicates that people are much more worried about radiation from nuclear power stations than from 'natural' radon. If the object is to reduce radiation exposure, it would be better to spend the available resources to reduce radon exposure in homes.

In the USA, Prof. Bernard Cohen has remarked that "In the past ten years science has come under irrational attacks from the forces of ignorance, and is losing public support. Our Government's Science and Technology policy is now governed by uninformed and emotion-driven public opinion rather than by sound scientific advice". It is not the actual risk, but the public perception of the risk, that determines the action taken.

The proposal in 1981 by the CEGB to build a nuclear power station at Sizewell in Essex was the subject of an extremely lengthy and costly public enquiry presided over by Sir Frank Layfield. Environmental groups and other opponents were given every opportunity to present their cases, and they were answered in detail by the CEGB. The enquiry took four years and the transcript of the proceedings ran to sixteen million words. Sir Frank Layfield's report was published two years later in 1987. He concluded that there was "good confidence that Sizewell B, if built, would be sufficiently safe to be tolerable, provided that there is expected to be economic benefit sufficient to justify the risks incurred. ... My conclusion is that Sizewell B is likely to be the least cost choice for new generating capacity". If one is to have such enquiries, the conclusions should be accepted, and indeed there was general acceptance of the findings. However the opposition groups unanimously condemned the report, and the Labour and SDP/Liberal parties said that they would cancel the project if they came to power in the election that was then due. However, Sizewell B was built, and is now fully operational.

Environmentalists continually call for public enquiries, and ensure by a series of delaying tactics that they take as long as possible. When finally the report of the enquiry is published it is immediately rejected, and a new enquiry demanded. This reveals that the real objective is to prevent anything being done. The aim is to stop the development of nuclear power, which they see as the supreme symbol and embodiment of modern technological society. In the words of the spokesman of the SDP study group on energy, "the nuclear threat is the hardest test the environmental movement has yet faced. Nuclear power lies at the heart of a vision of the future committed to an expansion of the present pattern of economic and industrial development. If we wish to argue for alternative patterns of economic and industrial activity, more rationally secure, more personally satisfying and more environmentally sustainable, then we must succeed in stopping the development of nuclear power".

It is widely believed, especially by the nuclear industry, that what is needed is more public information to counteract what is seen as a failure in communication. This may have been true in the past, but now the public is deluged with glossy brochures, fact sheets and magazines. The trouble is that people identify this as a public relations exercise and reject whatever is not in accord with their ingrained prejudices. Unlimited, accurate information is powerless against closed minds. An environmental spokesman has suggested that it is a waste of time to try to influence public attitudes by technical facts and figures, and that "decision makers should pay more attention to public perceptions of risks than to the actual risks". In other words, we should build our future on fantasy and not on fact. Reality, however, is not influenced by our fantasies and is liable to exact a terrible revenge on whoever treats it lightly.

It is important not only to respect facts, but to restore trust in the nuclear industry. Trust takes a long time to build but can easily be destroyed. In the case of nuclear power, trust has been destroyed by the media continually harping on radiation hazards, reactor accidents, radioactive waste and reactor decommissioning. It will take a long time to rebuild the trust that is necessary to secure our future.

8.7   Organising the Technological Future

Modern technology benefits us all on such a scale that there is no possibility of turning back to a simple lifestyle. Isolated individuals may retreat into the wilderness, but even they usually take supplies with them and rely on civilisation when a crisis occurs. Most of us do not have even this possibility.

Technology does, however, bring with it many serious hazards. Some of them, such as shipwrecks, are familiar, but others such as air disasters have become important only in recent years. We do all we can to reduce the incidence of such disasters, but they cannot be entirely eliminated. When they do occur they are accepted, and perhaps a commission of enquiry is set up to ensure that it cannot happen again. It is notable, however, that they are not accompanied by demands that the whole technology be abandoned.

There are other disasters characteristic of modern technology that are extremely serious because their effect are very widespread. Chernobyl is a prime example, but also very serious were the chemical disasters at Seveso and Bhopal.

Modern technological society relies on highly complicated devices such as aeroplanes and nuclear reactors that are made and operated by large numbers of people. They require a huge investment in manpower, finance and technical expertise, and the paramount importance of safety is clearly recognised. The designs are carefully checked and all components rigorously tested. At every stage in the construction the possibilities of failure are evaluated and reduced to negligible proportions. And yet, in spite of all this, catastrophic accidents still occur.

Sometimes this is due to hitherto unknown physical effects that only appear when a full-scale plant is operated for the first time. Examples are the Wigner release that caused the fire at Calder Hall, and the xenon poisoning that nearly stopped the operation of the first plutonium-producing reactors at Hanford. Neither of these were major disasters, and once they were understood they could easily be avoided in the future.

A much more serious cause of major disasters is the strong political and psychological pressures that are frequently present in large projects. These can lead to corners being cut, and components and procedures being approved without proper checks. In large organisations the responsibility can be so diffused that it is not clear who is really responsible for the vital decisions, and sometimes these are effectively taken by people who lack the necessary technical knowledge. There is no one person responsible, and so each individual can rather easily feel that it is not his or her responsibility. There may well be strong corporate pressure to keep to a tight timetable that can override better judgement.

The Chernobyl disaster has already been discussed, and shows how political pressures can impose a basically unsafe design, and inadequate operating procedures allowed the reactor to be controlled by men unfamiliar with the hazards. Operated according to the instructions laid down to avoid the dangers inherent in the design, it should have been safe, and indeed it was safely run for many years. Then, in order to carry out an ill-conceived experiment, the safety devices were switched off and disaster followed.

The Challenger disaster is even more instructive because of the much greater complexity of the space shuttle and the whole NASA organisation.

At 11.38 on Tuesday 28 January 1986 the Challenger orbital space craft with a crew of seven blasted off from Cape Kennedy. Seventy-three seconds later it disintegrated with the loss of all the crew members. The launch was watched by millions on the ground and on television. President Reagan was due to give his State of the Union address that day, and it was planned that during it he would talk to the astronauts. Instead, he broadcast a tribute to their courage.

The nation was shocked. How could a launch that was so carefully planned by the vast NASA organisation, supported by thousands of scientists and engineers, go so badly wrong? Were there not detailed and extensive safety checks at each step of the long process of design, planning and assembly of the spacecraft? Was it not agreed by all that it was a safe as possible, so that a civilian could be included among the astronauts?

This was not the first time that an engineering enterprise costing tens or hundreds of millions of dollars has ended in catastrophe. Another example is the Hubble space telescope. It was initially hoped that this instrument, a large telescope lifted into orbit around the earth would, because of the absence of distortions due to the earth's atmosphere, produce pictures of stellar phenomena of unprecedented sharpness. The whole project was extremely costly and was designed by a large team of scientists and engineers. And yet when the first pictures were received, they were found to be quite fuzzy. Investigations soon showed that the large mirror, so carefully ground and polished, suffered from spherical aberration. Fortunately it proved possible to design a correcting device that was inserted on a second launch, but at an extra cost of forty million dollars.

Similar stories could be told of many air disasters. An early one, now almost forgotten, was the fate of the R101 airship on the way from London to India. It took off in rough weather and crashed in France. On board were the members of a top Government delegation, and they carried with them heavy carpets and other presents.

Most, if not all, of these tragedies need never have happened. What went wrong, and how can they be avoided in the future? Commissions of Enquiry were appointed, and in such circumstances there is a reluctance to assign blame; this will not bring back the dead. It is all too easy to issue a bland report and to call it an accident. But they were not accidents; they need not have happened and should not have happened.

The Commission of Enquiry into the Challenger disaster included Richard Feynman, a scientist of legendary genius, who was determined to get to the bottom of the disaster. Pictures of the rocket soon after take-off showed a puff of gas from the joint between two sections. These are sealed by rubber O-rings, and it was suggested that their failure might be the cause. During a televised session of the Commission, a sample O-ring was passed around for inspection. Feynman put the O-ring in a clamp and immersed it in a glass of iced water. When it was cold, he took it out, removed the clamp and showed that the rubber did not spring back to its usual shape. The launch took place in unusually cold weather, and it was clear that this was most likely to be the cause. It was known to the manufacturers that the resilience of the O-rings decreased at low temperature. Feynman wrote later: "On the night before the launch, they told NASA that the shuttle should not fly if the temperature was below 53 degrees — the previous lowest temperature — and on that morning it was 29".

That investigation established the technical cause, but Feynman insisted that it was also necessary to find out how it happened. It soon became clear that there were serious defects in the whole organisation. The leaders know that they are dependent on Congressional funds so they painted a glowing picture of the successes achieved and the plans for the future. Much further down the organisation, the scientists and engineers were increasingly concerned that the technology was being pushed up to and perhaps even beyond the limits of safety. They wanted to make further studies and tests but all this takes time and money. There was strong political pressure to keep to a probably unrealistic schedule. Launches can only take place at certain particular times, and if a favourable opportunity is lost, the whole space programme is set back. This can force the scientists and engineers to relax the controls, to omit tests and to take short cuts. It is not easy for a scientist or engineer to voice his anxieties; his evidence may only be a matter of probability and indeed may be wrong. His superiors tend to be unreceptive to bad news, and if it came to the notice of the Press the publicity would be extremely damaging. And so gradually the tragedy unfolds.

Feynman was adamant that the disaster did not have to happen; it was not an accident. He suggested that there is a beguiling but deadly argument at work. Once several launches are successful, then it is safe to make more if the conditions are not so different. It is in the range of 'safe' parameters. He compared this to the case of a child who runs across the road in defiance of his mother's warnings. He can run across several times without being hit, but this does not prove that it is safe to do so. Sooner or later the child will be hit, so it is not an accident and it should never have happened. He concluded: "Officials must live in a world of reality in understanding technological weaknesses. For a successful technology, reality must take precedence over public relations, for nature cannot be fooled".

The need for extreme care is always necessary when a new technology is being developed, and airships are no exception. During the design of an earlier airship, the R38, no calculations were made of the stresses due to the aerodynamic forces to which the airship would be subjected. After it crashed, it was found that information about these forces existed, and consultation would have revealed that the safety factors were sometimes under two and in conditions that might easily occur could even be less than one. The R101 was much more carefully designed, but there was not enough time for a complete re-examination of all the aerodynamic calculations required by the modifications to the original design. After the crash the investigating commission found that it was probably due to loss of gas, which could easily happen in the rough weather at that time. The flight should have been postponed for a proper programme of tests but it was made, despite the misgivings of the engineers, because political considerations determined the date of the flight to India. Once again, as in the Challenger disaster, political considerations overrode the technical ones.

The mirror for the Hubble telescope was ground to exacting specifications by a sub-contractor. During the testing process a fault developed in the test bench, but it was assumed that this was not important. Some tests indeed showed evidence of spherical aberration, but they were not taken seriously.

The moral of these tragic events is clear. In any large organisation it is essential to keep the lines of communication clear, so that any reasonable doubts can be rectified in time. An independent body within the organisation could be the recipient of any doubts, and it could be given the authority to take the necessary action. The Nuclear Safety Inspectorate plays this role in the British nuclear power programme. They have the power to go everywhere and to examine everything, and they can shut down a reactor immediately if they have any doubts concerning its safety.

It is not an easy matter to say what is a reasonable doubt. It is always easy to say that one more test is necessary, but then there is no end. At some point it is necessary to make a decision, and that is best left to the professional judgment of the engineers. When they have made a decision, it is essential that they be taken seriously, and not overridden by political and other pressures. These can be extremely strong, and are both implicit and explicit. The implicit pressures are the fear of being regarded as a trouble maker, of losing promotion or employment if one stands up against the strongly expressed wishes of the management. The external pressures are budgetary constraints, commercial deadlines and Government directives.

It is only the major disasters that hit the headlines, but in any organisation, in business, industry, schools and universities, there are minor disasters that could easily have been foreseen and prevented. It is often evident that a policy, adopted because of financial or political pressures, will in the end prove disastrous. The laws of nature are inexorable, and if they are flouted they take their terrible revenge. Disasters generally happen, not by just one spectacular error, but by a whole series of minor compromises, cover-ups, organisational defects and false economies. The price of safety is eternal vigilance.

Reference

Rotblat, J., Reminiscences on the 40th Anniversary of the Russell-Einstein Manifesto, *Pugwash Newsletter 33*, No. 1/2, July/October 1995, p. 48.

9

MORAL ASPECTS OF THE ENERGY CRISIS

9.1 Responsibility for the Earth

Basic to the whole question of energy and the environment is our responsibility for the earth. Are we entitled to use the resources of the earth just as we please, or are there some principles we should follow? Does the earth have absolute rights, or are they relative to those of mankind? Should we just use the earth to satisfy our needs? And what counts as a need? May we also use it for our amusement? Should we be content with just picking nuts and berries, and digging up edible roots, or may we kill animals for food, replace forests by farmland, dam rivers, irrigate plains, mine coal and other minerals, sink oil wells and so build a technological society? Is there any limit to this process? Intimately connected with these questions are those of our responsibilities to each other. Some people live on rich land with plentiful energy supplies while others are crowded together in poverty on poor land. Are the more fortunate justified in living well and then selling what they do not need to poor people who desperately need it but cannot afford to pay for it? The rich become even richer while the poor die or sink further into debt. Do the rich have an obligation to share their wealth, and if so how far does this obligation extend? Just to their own families, or to people in the same town or country, or to everyone on earth? And should we not think not only about those living now, but also future generations?

When they tackle their energy problems, countries cannot act alone. Often they need essential fuel from other countries, and the way they generate their energy affects other countries not only indirectly by its economic repercussions but directly by the pollution that is no respecter of frontiers.

Many people believe, or act as if they believe, that all economic matters should be decided by market forces alone. They just let market forces determine what happens. What this means is that greedy and powerful people take far more than they need, leaving the rest in poverty. They do not care if their power plants or factories pollute the rivers and the atmosphere so long as they can maximise their profits.

Most people know that this is simply wrong. The great religions of the world teach that we should respect the natural world. The Judeo-Christian tradition, Islam, Buddhism, Hinduism and Shinto, to mention just a few, all emphasise respect for nature.

Many early religions indeed made no distinction between man and nature. We are surrounded by forces, some evil and some good, over which we have little control. Early man personified these forces as gods to be won over with offerings and sacrifices. In such religions it would have been unthinkable for man to try to control nature and tame it for his own use. In any case he lacked the knowledge to do this.

It is often said that primitive man walked lightly on the earth, taking only the plants and killing only the animals to satisfy his immediate needs. The bushmen of the Kalahari desert in South Africa have no concept of ownership, and share all they have with other bushmen. The speech by Chief Seattle of the Duwamish Indians in 1855 has been widely quoted. Responding to an offer from the US President to buy his land, he replied: "How can you buy or sell the sky? The land? The idea is strange to us. Every part of the earth is sacred to my people...The Earth does not belong to man, man belongs to the earth. All things are connected like the blood that unites us all. Man does not weave the web of life, he is merely a strand in it". These moving words were however written by Ted Perry in 1971 for a fictional television programme. The only record of what Chief Seattle actually said in 1855 was first written down thirty years later, and according to this the Chief simply praised the generosity of the President for buying his land. Mr Perry never claimed to be writing anything but fiction, but his words have been widely used to perpetuate the myth of the noble savage. Primitive peoples, even more than ourselves, were fighting for survival for most of their lives, and did not hesitate to take what they needed, often causing great environmental damage. The Polynesians killed the moas of New Zealand to extinction. They may have done less harm than people in the twentieth century because their power of inflicting damage was so much less, but on the whole their attitude to nature was much the same as ours.

The clear distinction between God and nature was first made by the Hebrews about three thousand years ago. They believed in one God, the supreme spirit who alone created the world and keeps it continually in being. Without God the universe would never have existed, and without His continual sustaining power it would immediately lapse into nothing.

The second sharp distinction made by the Hebrews concerns man. Man is a part of nature, completely dependent on God for his whole being, and yet set apart from nature in an absolutely unique way. Unlike the animals and plants, we have personal freedom. We can decide what to do; we can choose between good and evil. The special dignity of man is emphasised in the creation story in the Bible. Man is created last of all, when the rest of nature has been prepared to receive him, and he is given a special responsibility for the rest of nature: "Be fruitful, multiply, fill the earth and conquer it. Be masters of the fish of the sea, the birds of heaven and all living animals on the earth (Gen 1.28). God took man and settled him in the garden of Eden to cultivate and take care of it (Gen 2.15). The Psalmist praised the Creator who made man "little less than a god, and has crowned him with glory and splendour, made him lord over the work of his hands, and set all things under his feet". Man is told not to "bring destruction by the works of your hands, because God did not make death, and he does not delight in the death of the living" (Wisdom 1.12).

Today this teaching is largely ignored, as the earth and the seas are plundered to supply the ever-increasing demands of a consumer society. The forests are destroyed, the earth and the seas polluted and whole eco-systems threatened with destruction. "The earth is defiled under its inhabitants' feet, for they have transgressed the law, violated the precept, broken the everlasting covenant. So a curse consumes the earth, and its inhabitants suffer the penalty". (Isaiah 24.5)

The harmony between man and nature will not be restored until we have seen the grave damage we are doing, and have resolved in future to respect the whole of creation and to live in a sustainable way. It is not sufficient to make fine speeches and to pass resolutions professing respect for nature; it is essential to study the situation in all its complexity and to implement effective actions. This means examining world manufacturing policies and deciding what changes need to be made. A vital decision is what type of power stations to build and how to ensure that they are as safe and environmentally friendly as possible.

It is sometimes suggested that the energy crisis can be solved by reducing world population. While it is certainly necessary to prevent the world population increasing without limit, this by itself is not only a superficial way to try to solve the energy crisis, but is inadequate on its own. It is also important to examine our lifestyles to see if we could behave in a more civilised way. At present we use vast resources to manufacture military equipment and a large number of other machines that will soon be destroyed or become obsolete. Artificial and even unhealthy needs are artificially created so that more goods have to be made to satisfy them. Such a society is insanely wasteful, squandering the natural resources of the earth and destroying its beauty. Such behaviour is neither intelligent nor civilised. As long as this situation endures, we should not say that there are too many people on the earth.

Most people in the developed countries have incomes that are many times what is needed for subsistence. We are then responsible for the suffering that could be avoided if we had used the excess to relieve that suffering.

James Lovelock, in his books on Gaia, has shown that the animate and inanimate processes of the earth form a tightly interlocked system, with complex feedback mechanisms that keep it in a stable state. If however it is pushed too far, it may change irreversibly to a new stable state. Many environmentalists have used this to elevate the earth to the status of a living system — the earth-mother who look after us all. Lovelock has been elevated to the status of a guru.

Lovelock, however, is a practical scientist, not a woolly idealist, and is critical of the failure of the environmentalists to protest against the real threats to the earth such as bad farming techniques, and instead waste their time opposing nuclear power. He recognises that nuclear radiation is a natural part of the environment, and that God's universe, the sun and the stars, is basically nuclear. Concerning the effects of radiation, in particular the fear of cancer, he noted that there are many ways that the living cell can be damaged, by oxidation and other chemical processes. The cancer deaths from coal power should be our serious concern.

There was even a nuclear reactor at Oklo in the Gabon about 1800 million years ago that ran at the level of kilowatts for millions of years in a particularly rich vein of uranium ore, moderated by ground water. The fission products from this reactor moved hardly at all in millions of years. There is now no trace of plutonium and the 235 U is significantly depleted, so the uranium deposit is less radioactive than it would have been if the reaction had not taken place.

9.2 Practical Applications

What does this mean in practice? In primitive societies, areas of the forest are cleared for cultivation, and when the soil is exhausted the tribe moves to another site. It is when people settle permanently, each with his own land, that the problems arise. Careful farmers soon learn that to ensure continued fertility it is necessary to return to the land what is taken from it, to rotate the crops, to use fertilisers and, depending on the quality of the land, to let it lie fallow periodically.

In our modern technological society we need to apply to the whole earth the same habits of careful husbandry. The problems are however immensely more complicated, not only because the economies of distant countries are interlinked, but also because industrial developments lead to pollution of the environment. It is no longer just a matter of each farmer caring for his own land; the responsibility for the earth extends also to the whole community at the national and at the international level.

To ensure the care of the earth, regulations governing the use of materials and the control of pollution have been established and enforced. The difficulty is that pollution control, like safety, costs money, and could, if adopted, easily render an industrial process uneconomic. There are thus strong economic pressures to reduce it to the minimum, or avoid it altogether. However well-intentioned they may be, companies cannot adopt measures that will put them out of business. This difficulty is avoided if national regulations impose the same rules on all companies. In many cases this applies on an international scale, and so international regulations are necessary. It is not easy to establish the detailed regulations saying exactly what is permitted and what is not, or to ensure that the regulations are obeyed. It is very easy to say that everything must be made perfectly safe, and all pollution must be forbidden. Unfortunately this is quite impossible. Everything we do carries some element of risk, and most industrial processes are polluting to some extent. It is usually relatively easy to make some improvements, but the more are made, the more it is difficult to make further improvements. Thus, for example, the first half of atmospheric pollution by coal power stations is relatively easy to remove by filters, but the last 5% is far more costly. Thus a delicate balance between the cost of the anti-pollution measure and the social cost of the pollution has to be struck before a wise decision can be made.

To achieve this, it is essential to have detailed knowledge of the industrial processes and of the effects of the pollution they produce. Even if this knowledge is available, the pressure of ill-informed public opinion often forces companies and Governments to take what are objectively the wrong decisions. The public appreciation of relative risks is often quite wrong; there is paranoic anxiety about relatively minor risks, while major risks go unrecognised.

The two aspects of caring for the earth are the conservation of its resources and the avoidance of waste. To ensure this, manufactured objects should be made to last as long as possible. The manufacturer, however, wants them to wear out as soon as possible, so that we have to buy again. The environmental and commercial interests are thus directly opposed. There are many workmen who take a pride in their work and make things as carefully as possible, with good materials, so that they last a long time. The pressure on manufacturers are however so strong that it is now in their interests to make things so that they wear out quickly. This is clearly evil.

The market is now saturated with some types of electronic equipment. They are very reliable, so the manufacturers have to adopt another method. They announce that they have made a new model and after a certain date they will no longer maintain the old one, or only for an exorbitant sum. The user is thus forced to buy the new model, even if he is perfectly satisfied with the older one.

It is clearly desirable that these practices should be made illegal. Governments, however, are subject to political pressures, and will be told that such laws would curb production and put people out of work. The politicians may even be influenced in other ways by the manufacturers.

9.3 The Problems of Developing Countries

The developing countries are the ones that are hardest hit by the energy crisis. Most of them have no oil or coal, and so they are heavily dependent on imported oil for their energy supplies, their industries and their domestic lighting and heating. If the price of oil rises they are immediately in a very difficult situation. They are usually already heavily in debt and there is very little scope for reducing unessential energy consumption to protect essential services. They have no alternative but to reduce oil imports or plunge more heavily into debt, and often both at once.

The poorer countries often export fruit to pay for their imports, but the terms of world trade often change to their disadvantage. For example, before 1973, Costa Rica had to export 28 kg of bananas to pay for one barrel of oil. Ten years later one barrel of oil cost 420 kg of bananas. Already the budgets of many African, Asian and South American countries are overstretched by the need to buy oil for essential purposes. The price of oil fluctuates, but on the whole it tends to increase because the oil-producing countries want to maximise their profits and because the richer countries are using more than their fair share.

In the Third World, wood is still extensively used, and if oil is too expensive the increasing demand for fuel leads to the destruction of the trees and the creation of a wasteland. The country for miles around many African towns is being stripped of anything that can burn. As soon as saplings are planted, they are ripped up for fuel. The children, instead of attending school, spend hours searching for sticks and twigs. Animal dung, that should be returned to the soil, is dried and burnt as fuel. The productivity of the land inevitably falls, leading to a downward spiral of poverty, malnutrition, famine and death.

Such people cannot be helped by providing power stations, because in the rural areas and often in the towns they are not electrified. They cannot afford the wiring and the electrical equipment. What they need is cheap oil, and world oil prices are likely to come down if the developed countries use less oil, for example by building nuclear instead of oil power stations. As their economies develop, the poorer countries will be able to afford electrification first in the towns and then in the villages, and at this stage they will need power stations. It is particularly desirable that relatively small power stations be developed.

There are already many large towns and cities in even the poorest countries, and their energy needs cannot be met by burning wood and coal without severe pollution. People move from rural to urban areas and each person needs more energy. Some of the largest and fastest-growing cities in the world are in relatively poor countries. The population of Mexico City is expected to reach thirty million by the year 2000, and Bombay and Calcutta twenty million each. There will be about ten more with more than ten million people, and four hundred with more than a million people. Already many of these cities are heavily polluted, so how can they be given the energy they need without increasing this pollution to an intolerable level?

The present world situation is extremely complicated. Developed countries in Europe and North America generally have many possible energy sources, and are rich enough to buy what they need from other countries. They are able to enforce pollution controls and on the whole the state of their environments are improving year by year. Other countries, such as parts of Eastern Europe and

Russia have severe environmental problems due to decades without effective controls, and also some problems about ensuring a reliable supply of energy. Many if not most of the developing countries have a severe energy shortage and environmental problems ranging from erosion and desertification in rural areas and atmospheric and chemical pollution in the cities. So many people in the developing countries lack sufficient energy that there is an overall world energy shortage that requires urgent attention. Since each country has different needs and resources, it must have its own energy policy, and this must be co-ordinated with those of neighbouring countries.

The developing countries vary enormously among themselves. Some are small and poor, with little more than subsistence agriculture. Others like India have substantial industries and the expertise to construct and run their own coal, oil and nuclear power stations. Bombay, for example, has had a nuclear power station for some years. But all these countries, to a greater or lesser extent, find it impossible to cope with the problems of the present and to build the industries needed for the future.

What can be done about this? The poorer countries cannot afford to establish the industries that they need to earn the money to buy oil. If they are fortunate enough to have their own coal they are able to supply their energy needs, but at the cost of pollution. Nuclear power is certainly able to supply the energy needed by the large cities, without adding to the pollution. It would indeed reduce the pollution because if cheap electricity were available it would no longer be necessary to use wood and dung as fuel, which are the principal sources of pollution.

Unfortunately few of the developing countries can afford to buy nuclear power stations, so some means of providing them cheaply must be found. If they are lent the money needed they will run up huge debts that they will have little hope of repaying, and this leads to the richer countries dominating the poorer ones. They can use their power to force the poorer countries to buy their own goods and accept the prices they set for the material they buy from them. This is called economic imperialism.

In a speech to a conference on nuclear power held in Vienna in 1982, the leader of the Vatican delegation emphasised that the development of nuclear power is essential for the poorer countries of the world, and called for its general acceptance as a new source of energy. He said that to help the poorer countries a fund should be established to enable them to buy the machines and power stations that they must have if they are ever to escape from their poverty.

There are many serious difficulties, both economic and political. The economic ones could perhaps be overcome by permitting an international company to build a power station in a needy country at its own expense, and then sell the power produced at an economic rate. Thus the Westinghouse company has built a 620 MWe nuclear power station at Bataan in the Philippines for an initial payment of $100 M and will operate it for thirty years for $40 M per year and about 2.90 p/kW for the electricity produced. This supplies electricity to Manila, a city of six million people, and this could not be obtained in any other way without serious pollution.

There would still remain the fear in the company that after the power station is built, it would be taken over by the Government of the country. When this is done, the reactor could be used to produce fissile material for weapons; this is the problem of nuclear proliferation. This could perhaps be prevented by deeming the land occupied by the power station to be United Nations land, and guarding it by UN troops. Given the desperate need for energy in the poorer countries, it should be possible to overcome the economic and political problems of providing it for them.

Some of the most serious pollution occurs in the developing countries, and with it health hazards and the destruction of the environment. The most heavily polluted cities are to be found in these countries. If a country is in a desperate economic situation, and its industries are struggling for survival, they have neither the will nor the resources to worry about safety. Since pollution is no respecter of frontiers, this is a problem for the whole world. Inevitably, any attempt to force pollution regulations on a developing country is liable to run into strong opposition. It may easily happen that the installation of cleansing equipment to reduce pollution, or the adoption of new safety devices, would be so costly that the company could not survive. It is then easy for the developing country to say to the rich industrialised nations that it is all very well for them to adopt a high moral tone about pollution and safety, because they know that it would put them at a commercial advantage relative to the industries in the developing countries. The richer countries developed their industries without worrying about pollution, so why should the developing countries not do the same. This is certainly a serious problem that can only be tackled at the highest international level. It is right to require high standards of pollution control and safety, but this cannot be at the price of bankrupting the industry concerned. This could be overcome by subsidies designed to offset the losses incurred by following the international regulations.

These problems are so complex and inter-related that it is extremely difficult to see how they can be tackled successfully. It is well-known that countries hardly ever act altruistically, especially when their vital interests are at stake. It is no good just exhorting the richer countries to reduce their living standards and their high levels of consumption, or the poorer countries to avoid polluting the atmosphere. The force of public opinion puts severe constraints on the decisions of politicians, and in most countries it is not electorally popular to increase foreign aid beyond a token amount. It is practically impossible for governments to impose pollution controls on their own industries that will have the effect of putting them at a severe disadvantage compared with similar industries elsewhere.

To have any chance of success, these problems must be tackled at the highest international level, that is by the United Nations. Only such a body can impose energy policies and pollution controls that have any chance of being obeyed. If all countries agree to impose the same controls, then their industries will be affected in a similar way so that no one gains an advantage over the others. Even then it may be easier to follow the new rules in one country than another because, for example, the indigenous coal in one country has fewer impurities than that in another country. To allow for such differences it may be necessary to arrange appropriate subsidies. Already there is strong pressure to reduce the emission of carbon dioxide to agreed levels. This could be enforced by imposition of a carbon tax payable by every country according to the level of emission. This could be extended to form part of a wider network of pollution controls. The income from these taxes could be used to assist the poorer countries to modify their factories and power stations to reduce the level of pollution.

The richer countries could also tax luxury activities that add to pollution and in this way set an example that would make it easier for the poorer countries to accept the necessary controls. All such measures will inevitably be highly unpopular and it will not be easy to agree on the necessary regulations, but it is in the long-term interest of every country to take the necessary measures to preserve the earth for future generations.

At the same time as these high-level negotiations it is essential to do all that is possible to educate the peoples of all countries about the need to moderate their lifestyles, reduce pollution and in general to care for the earth. Already it is evident that there is a deep concern for the environment, especially among young people. The phenomenal support for the environmental movements shows the strength of this concern, although sadly the activities of some of these movements are counter-productive. As always, effective action requires not only concern and enthusiasm, but also awareness of the scientific and technical data that are the essential basis of any realistic policy. As people realise the need, they will more readily support environmental policy decisions.

These remarks concern what can be done to improve the situation in the poorer countries of the world. Whether it will be done depends on whether the will is there to effect the necessary changes. A country can be helped by aid from outside but whether it makes fruitful use of that aid or whether it is simply squandered depends on the determination of its people. If the will to improve is not there no amount of aid will be effective in raising living standards. There are many examples of countries, well-endowed with natural resources, that remain in poverty due to a corrupt political system. In such countries the wealth is concentrated in the hands of a small minority, and the vast majority of the population is illiterate, undernourished and over-taxed. The cities are polluted and filthy and crime is rife. Feudal families, a corrupt bureaucracy and the military rule the country. Most of the annual budget is spent on armaments and on servicing foreign debts, while only a minuscule fraction goes to health, welfare and education. Their universities are farcical imitations of those in developed countries, and it requires heroism to do serious work in them. Such countries have the means to improve their living standards, but they lack the will to do so.

At the other end of the scale, there are countries like Japan that are not particularly well-endowed with natural resources but have a strong determination to master all available technologies, to improve and develop them and to build a modern technological society. They have a highly-developed educational system, a high standard of personal morality and an awe-inspiring capacity for work. The transportation system is fast and highly efficient and the cities, though very crowded, are clean, pleasant and safe places to live. Scientific research is very well-supported, and contacts with scientists in other countries are numerous and well-funded. The result is a level of scientific and technological achievement that is one of the highest in the world. In a little over a century they have raised themselves from an isolated, backward feudal society to become one of the most highly developed technological countries in the world.

It is notable that the 'tiger' economies of South-East Asia, with economic growth rates of about 7% per year, are investing heavily in nuclear power.

Already Japan, Korea and Taiwan are 30 to 40% nuclear, and China, India and Pakistan have established nuclear programmes. Their energy consumption is expected to double by 2010, and accounts for 21% of the world's carbon dioxide emissions. This is expected to rise to 25% by 2010 and, without nuclear power, the figure could be much higher.

9.4 The Threat of Nuclear Weapons

One of the strongest arguments against nuclear power is that it requires the establishment of a nuclear industry that is able to build not only nuclear power stations but also nuclear bombs. In many respects the technology is very similar. The spent rods of uranium from a reactor can be processed to extract plutonium, and this can be made into bombs. Nuclear war is such a terrible threat to the life of mankind that we might even be willing to forego nuclear power if thereby we could remove the possibility of nuclear war.

Whatever we may think about this, it is not an option that is open to us. Nuclear reactors are already operating in many countries, supported in most cases by a nuclear industry. We cannot turn the clock back; nuclear power is here to stay. The danger of nuclear war already exists, and would hardly be affected if we were to demolish all existing reactors. Indeed it is arguable that such an action would actually increase the danger of nuclear war, as it would have the effect of exacerbating the energy crisis and thus strongly increasing the demand for oil. The resulting international tensions, as the nations scrambled for the remaining oil supplies, could easily be more serious than the threat due to the presently-existing reactors.

The technology of nuclear reactors is now well understood, and many nations have enough scientific and technological expertise to enable them to build a nuclear reactor able to produce weapons-grade plutonium. It is also possible to produce fissile uranium for bombs by separating the two uranium isotopes using a centrifuge. The manufacture of the bombs themselves is not easy, but the methods used are now widely published. To be an effective weapon it need not be particularly well made. A bomb that explodes prematurely may not be as efficient as those made by professionals, but it could still cause immense damage.

There is also the danger of fissile material being stolen by guerillas so that they can threaten Governments to force them to yield to their demands. A more serious danger is that unreliable regimes may obtain weapons-grade uranium or plutonium from the former Soviet Union.

It has been argued that the operation of reprocessing plants increases the amount of plutonium in the world, and thus the danger of nuclear proliferation. The plutonium can however itself be used to generate nuclear power, particularly in fast reactors, so it is not wasted.

The dangers of nuclear proliferation would be greatly reduced if all reactors were subject to strict international control. This was indeed proposed as long ago as 1946 by Senator Baruch on behalf of the United States, at a time when that country had a monopoly of nuclear weapons. He urged the creation of an International Atomic Development Authority which would be entrusted with the responsibility for all phases of the development and use of atomic energy from the raw material to the final applications. His plan was that all nuclear activities potentially dangerous to world security would be either owned by the Authority or subjected to effective control, and it would be empowered to control, license and inspect all other nuclear activities. It would also have the duty of fostering the beneficial uses of atomic energy, and of keeping in the forefront of research and development of new ideas and techniques. The United States offered to make all its accumulated experience and equipment available to the Authority. If this far-sighted plan had been accepted, the world would be a far safer place today. Unfortunately the plan was rejected by the Soviet Union, mainly because it seemed to them that it would leave nuclear weapons in the hands of the United States for an unknown period, while at the same time preventing any other nations from making them.

It is now too late for international control in the sense of the Baruch Plan to be a practical possibility, but nevertheless the idea of international inspection is still very relevant. There is always the danger that countries desperately short of energy will construct reactors without taking sufficient care, and this can be prevented if all reactors are subject to inspection and license by an impartial international authority such as the International Atomic Energy Agency in Vienna.

The danger still remains, and is very real today, that a country will agree to abide by the international regulations and proceed to build up a nuclear industry. It may then suddenly refuse to allow the international inspectors to enter the country, as has indeed happened in Iraq and in North Korea. Immediately the surrounding nations fear that the reason for this is that it has been decided that the hitherto peaceful nuclear industry is about to be used to manufacture bombs. In the case of Iraq this fear was well justified, for it was subsequently found that enriched uranium had been removed from a licensed reactor to produce weapons. Following this, the IAEA inspection procedures have been strengthened.

In the early years of the nuclear arms race, nuclear weapons were frequently tested, and many of the tests released large quantities of radioactivity into the atmosphere. The contamination was particularly heavy for the first hydrogen bombs tests at Bikini and Eniwetok, and the Japanese fishing boat Fukuru Maru was so heavily contaminated by fallout that the fishermen suffered severely from radiation sickness. There was thus strong international protest when the French announced that they would resume testing in the Pacific. In this case, however, the test explosions were underground, minimising the release of radioactivity into the atmosphere.

These are very real dangers, and there is little apart from diplomacy that can be done to remove them. These problems are however now separate from those of energy production except that, as already mentioned, world energy shortage is a potent source of international tension, which could be one of the reasons why a nation would wish to develop its nuclear weapons capacity.

9.5   Conclusion

During the writing of this book my views on the problems of energy and environment have changed and developed in many respects. Initially, I was most concerned about future energy supplies; will there be enough energy to maintain a reasonable standard of living for a rapidly-increasing world population? If the rate of increase of world population continues to fall, it now seems likely that it will eventually stabilise at about two or three times its present level. This will require much more energy, though the total need can be reduced significantly if we learn to moderate our life-style. Even so, the world's energy needs will increase to roughly five times the present level, taking into account the need to increase the living standards of people in the poorer countries. It now seems very likely that an energy demand of this magnitude can eventually be met if the necessary steps are taken in time.

What is far less certain is whether this energy, and all the manufacturing industries necessary to increase world living standards, can be provided without causing irreversible and unacceptable damage to the environment. If we persist in pouring noxious chemicals into the earth, the atmosphere and the seas, then gradually and inexorably, the trees and the plants, the animals and the fishes, will die, and our health and quality of life will go with them. It is therefore imperative to examine and control noxious emissions from all industrial processes, including power production. The most polluting sources are those that burn fossil fuels, namely coal, oil and gas, and these must be phased out as soon as possible. The alternatives are the renewable sources, namely, hydroelectric and wind, together with nuclear. Hydro and wind also pollute the environment in different ways, and have inherent limitations.

During normal operations, nuclear power stations are relatively non-polluting, but they contain very large quantities of highly dangerous radio-activity. The techniques necessary to contain this radioactivity so that it poses no danger are very well understood, so the question to be faced is whether we have sufficient political will to enforce the necessary controls and to maintain eternal vigilance. This is no more than is required in any industry or transport system that uses high technology, but in the case of nuclear power the potential for disaster is greater than in most other industries, though not all.

If the answer to that question is yes, then we have a source of energy that is able to meet all reasonable needs for the foreseeable future in a relatively safe way that causes rather little harm to the environment. This would require the construction of more thermal reactors, with the addition of fast reactors as soon as they become economically desirable. Such power stations would be able to satisfy the increasing world energy needs, and enable the polluting fossil fuel power stations to be phased out. At the same time, hydro and wind could supply useful supplementary amounts of energy even when environmental constraints are imposed.

If however the answer to the nuclear question is no, then we could try to make do with far less energy, and exhaust the remaining supplies of precious oil and gas more rapidly, and this will inevitably lower our standard of living and make it very improbable that the conditions of people in the poorer countries will ever be improved. Alternatively, we could generate as much energy as we need for hundreds of years by building more coal power stations, and inevitably be forced to endure the greatly increased levels of pollution.

Which path will be chosen depends on many factors, particularly the decisions of politicians, which are determined by public opinion. Here there is real ground for pessimism. The level of public understanding of the complex problems of energy and environment is extremely low. It is not just ignorance, which would be understandable, but the prevalence of strongly-held views that are contrary to reality by people who are not open to reasoned argument that is so dangerous. The ignorance is partly due to the need for the quantitative evaluation of all relevant factors, but still more to the activities of the mass media and many environmental organisations. While there are some sections of the mass media that do try to present a balanced view, far too often they are driven by their continual urge for scandal and sensation. Some environmental organisations do much good work, but like the media they are generally insensitive to quantitative considerations. Both frequently show a pathological aversion to nuclear power that is not affected by a detailed comparison between the various energy sources. This inevitably forms public opinion and hence forces politicians to take decisions that are not in the long-term public interest. An additional complication is that most energy policy decisions are rather long-term, so that plans have to be made over periods of several decades. Politicians tend to be concerned more with the next election, and are not so concerned about long-term energy planning. It is thus all too likely that the wrong decisions will be taken.

Another consideration that has become increasingly clear is the difficulty of obtaining accurate data on which sound decisions can be made. These difficulties are ones of principle, not just practical. Thus if we want to know the hazards of say coal power, we can collect statistics of the number of mining deaths and so on. These are however falling from year to year as improved safety measures are implemented, and also differ from one country to another. It is thus impossible to obtain more accurate figures. Further sources of difficulty are uncertain base-lines. Thus if we want to estimate the deaths from say cancer from an accident like that at Chernobyl we need to know the death rate both before and after the accident. However the records before the accident are very poor whereas those after are much more extensive and probably include many cases that have not been included before. If we attribute the difference between the two figures to the accident we will probably get an inflated result. Another difficulty is the likelihood of feed-back effects: after the accident there is an enhanced concern and knowledge about cancer and so it is likely to be detected earlier and so can be treated with a greater chance of success. In this way the accident can actually reduce the death rate.

The widespread fear of nuclear radiations has led to a very unbalanced treatment of various hazards. The fear that nuclear power stations and reprocessing plants are responsible for childhood leukaemia has forced British Nuclear Fuels to spend hundreds of millions of pounds to reduce the level of radioactive discharges, although it is already so low that it could not conceivably cause any detectable medical effects. This has diverted attention from the task of identifying the real cause of the additional cases observed. Generally speaking, radiation has far too prominent a place in discussions of optimum energy sources. Most dangers that are endlessly discussed are quite negligible compared with other risks we take daily. It is only in rare and exceptional cases that nuclear radiations pose a real problem.

It is sometimes said that power production must be made absolutely safe, but this is unfortunately impossible. We should of course strive to improve safety, but it is not always realised that there is a practical limit to this. The duplication of safety devices costs money that could be spent more efficiently elsewhere. Furthermore the production of safety devices itself involves hazards, so eventually the attempts to increase safety have the opposite effect.

It is very difficult to obtain reliable figures of the costs of the various energy sources, for the reasons described in Chap. 5. The costs depend not only on physical factors, which can be reasonably well estimated, but on political factors such as the legal and safety requirements, the rate of return required on the initial investments, and the estimated life of the power station. When these factors are estimated, it turns out that the uncertainties in costs of the major power sources, coal, oil and nuclear are often similar to or greater than the differences between them. The same applies to wind, with some reservations. This is the case when no account is taken of the hidden costs due to environmental pollution, which can be very large. The result is that pollution is the principal factor that should control our choice of energy source. This is an important result that short-circuits many of the detailed discussions of optimum energy sources.

References

Derrick, C., *The Delicate Creation: Towards a Theology of the Environment*, Tom Stacey, London, 1972.

Lovelock, J. E., *Gaia: A New Look at Life on Earth*, Oxford University Press, 1979, 1987, 1993; *The Ages of Gaia: A Biography of our Living Earth*, Oxford University Press, 1995.

Marsh, B., *Towards a Theology of Ecology*, University of Navarre, Pamplona, 1994.

Willrich, M., and Taylor, T. B., *Nuclear Theft: Risks and Safeguards*, Cambridge, Mass: Bollinger Publishing Co., 1974.

APPENDIX 1

ENERGY UNITS

The basic energy unit is the *erg*, defined as the work done by a force of 1 *dyne* moving a distance of 1 cm. The *dyne* is that force which, acting on a mass of 1 gm produces an acceleration of 1 cm per sec per sec. A *joule* (J) is 10 ergs, and is also defined as the kinetic energy of a mass of 1 kg moving at 1 metre per second. Since this is very small for practical purposes, large multiples of the joule are frequently used, particularly the *exajoule* (EJ) (10^{18} J), the *gigajoule* (GJ) (10^9 J) and the *megajoule* (MJ) (10^6 J). In the oil industry the unit is the *tonne of oil equivalent* (TOE). 1 EJ = 22.7 million TOE, or 1 TOE = 44 GJ. Also, 7.3 barrels = 12 tonnes. Rates of heat production are measured in *watts*. A *watt* is the rate of working of one joule per second. A *kilowatt* (kW) is 1000 watts, a *megawatt* (MW) is 10^6 watts, a *gigawatt* (GW) is 10^9 watts and a *terawatt* (TW) is 10^{12} watts. One kilowatt hour (kWh) is 3.6 MJ. 1 EJ per year is 32.2 GJ per second or 32.2 gigawatts. One barrel of oil per day is 50 TOE per year. An energy unit used in nuclear physics is the electron volt (eV). One eV is 1.6×10^{-19} J. A million electron volts (MeV) is 1.6×10^{-13} J. Each fission of a uranium nucleus releases 200 MeV = 3.2×10^{-11} J. One gram of uranium 235 undergoing fission releases 82,000 MJ.

A more detailed discussion is in J. Ramage, *Energy: A Guidebook*, Oxford 1997.

APPENDIX 2

RADIATION UNITS

There are many different ways of measuring the intensity of nuclear radiation. Some of those mentioned below are now obsolete, but they are included because they may still be found in earlier publications.

The *roentgen* is a measure of the ionisation produced in a tissue, and is such that one roentgen produces two billion ion pairs in a cubic centimetre of standard air. The number of atoms in a cubic centimetre is so large that a roentgen ionizes only one atom in ten billion. This unit was originally defined for X-rays and gamma rays, and later a similar unit, the *rad*, was defined for any ionizing radiation. The rad corresponds to the absorption of a hundred ergs of energy per gram. Since a roentgen delivers about 84 ergs per gram the two units are very roughly the same. More recently a new unit, the *gray*, has been introduced. This is defined as the radiation corresponding to an energy absorption of 1 joule per kilogram. Thus 1 gray is equivalent to 100 rad.

Another important unit is the *curie*, defined as the radioactivity of 1 gram of radium. This may be extended to other radioactive substances by defining a curie as the radioactivity of an amount of that substance that has the same number of disintegrations per second, thirty-seven billion, as a gram of radium. For most purposes this is an inconveniently large unit, so the *millicuries* and the *microcurie*, one thousandth and one millionth of a curie respectively, are often used instead. Similar subdivisions are made of the roentgen and the rad. For some purposes it is convenient to use the *becquerel*, defined as the activity corresponding to one disintegration per second.

Since some types of radiation cause more damage than others, a unit has been defined to provide a standard of comparison between them. This is the *relative biological effectiveness* (RBE), defined as the dose from 220 KeV

X-rays causing a specific effect divided by the dose from the radiation causing the same effect. To make it possible to define the relative effects of various radiations on man an *effective quality factor* Q is often used. This has the value unity for X-rays, gamma rays and beta rays, ten for neutrons and protons and twenty for alpha-particles and other multiply-charged particles.

When considering the effects of nuclear radiations on man, it is also necessary to include the different sensitivities of the different organs of the body. This is done by defining a unit, the *rem*, that is the product of the absorbed dose in rads, the effective quality factor and any other modifying factor. The rem can also be defined as the dose given by gamma radiation that transfers 100 ergs of energy to each gram of biological tissue; for any other types of radiation it is the amount that does the same amount of biological damage. A more recent unit, the *sievert*, has been defined as 100 rem. A millisievert (mSv) is thus 0.1 rem.

FURTHER READING

Edgar Boyes (Ed), "Shaping Tomorrow" (Home Mission Division of the Methodist Church, 1981). A survey of modern technology, with excellent sections on electronics and nuclear power.

Bernard L. Cohen, "Nuclear Science and Society" (Anchor Books, 1974). A clear and simple introduction to nuclear physics, nuclear power, nuclear radiations and the implications for mankind.

Bernard L. Cohen, "Before It's Too Late: A Scientists' Case for Nuclear Energy" (Plenum Press, 1983). A clear and detailed analysis, with particular attention to comparative risks and media distortions.

Alan Cottrell, "How Safe is Nuclear Energy?" (Heinemann, 1981). A clear and authoritative account of the safety of nuclear reactors.

Clarence Glacken, "Traces on the Rhodian Shore" (University of California Press, 1967). Describes the relations of man and nature through the ages.

Wolf Hafele (Ed), "Energy in a Finite World: A Global Systems Analysis" (Ballinger Publishing Company, 1981). An authoritative survey.

Peter Hodgson, "Our Nuclear Future?" (Marshall Pickering, 1982). An account of the energy crisis with a detailed consideration of all the major sources of energy and their associated risks.

Herbert Inhaber, "Risks of Energy Production" (1981). A detailed study of the various types of risk involved in all the major ways of producing energy.

John Passmore, "Man's Responsibility for Nature" (Duckworth, 1980). A philosophical study of beliefs about man's relation to nature.

AUTHOR INDEX

SUBJECT INDEX